DESIGN FOR THE UNEXPECTED

DESIGN FOR THE UNEXPECTED

FROM HOLONIC MANUFACTURING SYSTEMS TOWARDS A HUMANE MECHATRONICS SOCIETY

PAUL VALCKENAERS
Faculty of Engineering Technology, KU Leuven

HENDRIK VAN BRUSSEL
Faculty of Engineering Science, KU Leuven

ELSEVIER

Amsterdam • Boston • Heidelberg • London
New York • Oxford • Paris • San Diego
San Francisco • Singapore • Sydney • Tokyo
Butterworth-Heinemann is an imprint of Elsevier

B
H

Butterworth-Heinemann is an imprint of Elsevier
The Boulevard, Langford Lane, Kidlington, Oxford OX5 1GB, UK
225 Wyman Street, Waltham, MA 02451, USA

Library of Congress Cataloging-in-Publication Data
A catalog record for this book is available from the Library of Congress

British Library Cataloguing-in-Publication Data
A catalogue record for this book is available from the British Library

ISBN: 978-0-12-803662-4

For information on all Butterworth-Heinemann publications
visit our website at http://store.elsevier.com/

CONTENTS

ABOUT THE AUTHORS

PAUL VALCKENAERS

Dr Paul Valckenaers is a Master in Computer Science. His seminal doctoral thesis on Flexibility for Integrated Production Systems (1993) formed the basis for the ensuing extensive research on design for the unexpected at the PMA lab of KU Leuven.

Ever since, as a postdoctoral researcher, he has coordinated a coherent collection of projects that finally led to the underlying book *Design for the Unexpected*. He published extensively on the subject and gave many invited presentations and courses worldwide on all aspects of his research.

His present research interests are directed toward extending the design for the unexpected paradigm outside the manufacturing world, particularly for eHealth, intelligent traffic, smart grids, homes, and cities.

HENDRIK VAN BRUSSEL

Professor Hendrik Van Brussel is emeritus professor in mechatronics and automation at the Department of Mechanical Engineering, Division of Production Engineering, Machine Design and Automation of Katholieke Universiteit Leuven (KU Leuven), Belgium.

He was a pioneer in robotics research in Europe and an active promoter of the mechatronics idea as the new paradigm for machine design. He has published extensively on different aspects of robotics, mechatronics, and flexible automation.

Professor Van Brussel is a Fellow of SME and IEEE. He holds honorary doctor degrees from RWTH, Aachen, Germany, from "Politehnica" University' Bucharest, Romania, and from "Transilvania" University Brasov, Romania. He was President of CIRP (International Academy for Production Engineering) and of euspen (European Society for Precision Engineering and Nanotechnology). He is a Member of the Royal Flemish Academy of Belgium for Sciences and Arts, Foreign Member of the Royal Swedish Academy of Engineering Sciences (IVA), Foreign Member of the US National Academy of Engineering (NAE), and Honorary Member of the Hungarian Academy of Sciences.

PREFACE

In writing a book there are two preconditions that must be fulfilled: (i) there is a subject worth to be written about and (ii) the author(s) have the time to write the book.

This book describes a research endeavor that spans a period of almost 30 years. The genesis of the subject of this book goes back to Japan in the early 1990s, when MITI (Ministry of International Trade and Industry) launched the ambitious IMS (Intelligent Manufacturing System) program, aiming at defining the desired structure of the factory of the twenty-first century. The second author, as head of the PMA (Production engineering, Machine design and Automation) lab of KU Leuven, and the first author, who just received his doctoral degree with a thesis on "Flexibility for integrated production automation," realized that joining IMS was an excellent opportunity to develop their innovative ideas on increasing the flexibility of manufacturing systems, they already had developed since 1986, far beyond the then prevailing CIM (Computer Integrated Manufacturing) paradigm. We joined the IMS feasibility study on Holonic Manufacturing Systems (HMS) as one of the 32 international partners from all over the world. We could elaborate on the seeds that were present in the first author's doctoral thesis. (At PMA we already had a holonic assembly cell up and running in 1988, as shown in Figure 7.1 of this book!)

Ever since we have been able to further develop the original seeds into a full-fledged tree, financially supported by KU Leuven, federal (Belgian), and regional (Flemish) funding, EU funding under the successive framework programs, and even direct industrial funding. This varied funding structure resulted in fundamental research results that proved their usefulness in a wide variety of application domains, as shown in Chapter 7.

A work of such a broad scope can only be realized by a large team and extensive funding. We are particularly grateful to our enthusiastic research team that, over the 30-years span, substantially contributed to the edifice that this book describes, by doctoral theses, dedicated research contributions and technical assistance: Luc Bongaerts, Jo Wyns, Patrick Peeters, François Bonneville, François Cottrez, Herman Claus, Jan Thielemans, Constantin Zamfirescu, Indra Tanaya, Yuki Indrayadi, Hadeli, Tony Van Ginderachter, Paul Verstraete, Bart Saint Germain, Johan Philips, and Osman Ali. We are equally grateful to the different funding agencies, mentioned earlier.

As to the second precondition, we had to wait for the retirement of the second author to push forward our long overdue book-writing project. Until then, chasing for funding and publishing papers, next to teaching of course, absorbed most of the time of the director of a research lab of the size of PMA. We thank Elsevier for their efficiency to have the book swiftly published. Indeed, time is overripe to spread the message that design for the unexpected is the only way to go for the future.

The title of the book may seem somewhat enigmatic and general to the reader, but the subtitle clarifies much. We hope that after reading and digesting the book you will be convinced of the appropriateness of its title and of the vast potential the subject has, not only to control manufacturing systems but entire mechatronic societies in all walks of life.

Paul Valckenaers,
Hendrik Van Brussel
Leuven, Summer 2015

INTRODUCTION

ONCE UPON A TIME

Somewhere in the 1980s, the industrial automation community initiated the development of computer-integrated manufacturing (CIM) systems. These were *systems of systems* aiming to integrate automated workstations into fully automated factories. In fact, this community was designing and developing systems of systems before it became a popular topic in systems engineering. Unfortunately for industrial automation, the results were underwhelming.

Within this setting, our research was looking for the root causes of the above. What causes smaller systems, when integrated into a larger system of systems, to collide? What makes it so hard to undo whatever is causing these collisions? Which aspects of those difficulties are intrinsically inevitable? What can be done? Which properties of an application domain (preconditions) allow us to remedy this undesirable situation?

At the outset, our expectation was to discover intrinsic limitations leading to a conclusion that little could be done. This would nevertheless be valuable whenever developers will avoid attempting the impossible (as an analogy of the second law of thermodynamics versus the perpetuum mobile). In reality, the research findings revealed to be quite the opposite. Although there are significant limitations, it proved to be possible to design systems improving the present situation in a wide range of application domains.

DESIGN FOR THE UNEXPECTED

Because the investigations refrained from making assumptions about the system of systems, in which the smaller systems are integrated, this book is about the design of systems for the unexpected. Indeed, in order to develop smaller systems that avoid and/or manage collisions when integrated into a larger system of systems, designers may not impose arbitrary constraints or at least must make it easy to revise them. Basically, this book lifts low-and-late commitment to a scientific level – where low-and-late commitment may be combined with early and eager preparation.

As a consequence, the research results are not only applicable to the integration into a system of systems, but they also address the frustration when attempting to improve large complicated organizations where an effort in one location fails to satisfy expectations because neighboring systems block the improvement. They also address the issue of changing user requirements as soon as a new system is deployed. Overall, they allow to develop systems while facing uncertainty about the future requirements that they will need to cope with. And they reveal how to design systems that will not become future legacy systems in the negative connotations of this word.

PREDICTING THE UNEXPECTED

This book will outline and discuss in detail how to design systems without relying on expectations that may prove to be wrong in the future and, when design choices nonetheless are relying on expectations, how to keep it easy to revise these choices. A major discovery was a generic scheme to generate short-term forecasts without forfeiting this approach to cope with the unexpected.

In fact, it were two laypersons who, after listening to an explanation of this scheme, labeled this achievement as "predicting the unexpected," after which we adopted this as our "nom de guerre."

INFORMATION INFRASTRUCTURE AND SUITABLE APPLICATION DOMAINS

Design for the unexpected cannot be applied to any kind of system. First, it addresses the design of information systems in relation to a world of interest. Moreover, this world of interest, the application domain, needs to possess certain properties such as the following:

- It contains valuable resources such as machines, roads, ambulances, wind turbines, workers, nurses, rooms, trucks, ships, robots, parking space, etc.
- It comprises activities using those resources such as commuting employees, freight transportation, patient treatment, warehouse refrigeration, production, etc.
- Intelligent coordination of those activities brings sufficient added value to justify the development effort and operating expenses of the information system.
- It is possible to mirror those activities and resources in executable computer models.

Finally, these information systems are information infrastructures that offer a sound basis to implement case- and situation-specific information systems.

TOWARD A HUMANE AND RESPECTFUL MECHATRONIC SOCIETY

The above suggests that information systems – designed for the unexpected – may become part of everyday life wherever and whenever valuable resources are involved. In view of our dissatisfaction with current situations (e.g., traffic jams) and the potential benefits (e.g., prevent toxic combinations of medications), such a "mechatronic society" (see Chapter 6) is expected to emerge. Moreover, other technological developments such as the Internet of Things contribute to the likeliness of such a future.

In this respect, our research reveals that the social and people aspects must not be addressed as afterthoughts. In fact, acceptable social behaviors, induced by minimally intrusive social control mechanisms, have been introduced and investigated within our software prototypes. Also, behaviors normally associated with humans revealed to be useful within these prototypes (e.g., opportunism).

Furthermore, the research revealed the need and opportunity to initiate interdisciplinary research, involving social sciences and humanities. In particular, there are opportunities to empower persons (e.g., when this empowerment allows systems to deliver superior services). It also becomes possible to reconcile high service levels with privacy. And respectful and considerate social controls can be implemented in the information infrastructure. As they inhibit abusive behaviors, this allows for the generation of common goods; this is especially value adding when combined with the mechanisms to predict the unexpected.

Overall, this book presents the onset for the development of information infrastructures that scale significantly beyond the existing and penetrate deeply into our lives. And they will be durable as they adapt to future demands avoiding to become legacy. Today, the main topic on our research agenda is to create a mechatronic society that is empowering, considerate, and hospitable toward humans. Our research results are inviting in this respect.

CHAPTER ONE

Setting the Stage

Paul Valckenaers*, Hendrik Van Brussel**
*Faculty of Engineering Technology, KU Leuven
**Faculty of Engineering Science, KU Leuven

Before embarking in this book for an odyssey to find a generic control design methodology for mechatronic societies, taking care of the unexpected (i.e., design for the unexpected [D4U]), we make our point by means of an illustrative example from daily life. The selected application domain is mobility, which is familiar to almost every reader and has adequate levels of complexity as well as unsolved issues. The chapter first presents a scenario illustrating the benefits of a D4U solution. Next, the manner in which a D4U is created is discussed.

A SAMPLE SCENARIO

This scenario illustrates how D4U decentralized traffic coordination handles a disturbance proactively. Paul is driving from Leuven to the FP7-ICT4EE event in Brussels (marked A on the map, Figure 1.1). His navigation system has started executing virtual journeys to this event beforehand; actually from the moment the event was entered in Paul's electronic agenda. The virtual journey execution strategy, implemented in Paul's navigation system, accounts for how much time is left until the actual journey. Up until the time when Paul starts his journey, nothing special happened and the navigation system arranged a trip that avoids rush-hour traffic and limits congestion charges. For the purpose of this scenario, payments for time slots depend on the actual demand (i.e., only for highly solicited time/location slots).

When Paul is about halfway his journey, a collision of two cars on the Wetstraat suddenly reduces the road capacity by 30%. The intelligent module (IM) corresponding to the road segment containing this location observes this incident through the information it receives from the onboard car navigation/safety systems and mobile phone tracking services. The IM is the first to flash a warning on the display wall of the emergency services, which investigate and indicate the severity and type of the incident on their system, to which the IM has subscribed by now and it improves its estimate

Design for the Unexpected. http://dx.doi.org/10.1016/B978-0-12-803662-4.00001-1

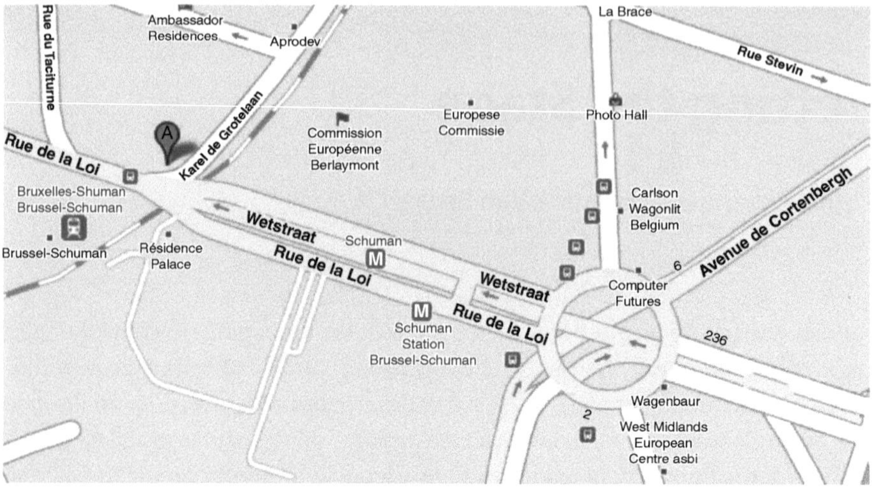

Figure 1.1 ***Location of the Unexpected Event in the Sample Scenario.*** *(Map data ©2011 Google).*

of the duration of this incident. The IM will keep updating this estimate as information becomes available.

When Paul's navigation system refreshes its journey registration, the impact of the incident becomes known to Paul. Instability suppression mechanisms prevent Paul's navigation system from overreacting too fast. At first, it starts to refresh the collection of alternative journeys that was kept up-to-date in case of these events. Alternatives that are also affected by the incident become equally unattractive. The navigation system also increases the intensity of its search for alternatives. Paul's navigation system analyzes the result of the refresh and discovers that the degraded performance has its root cause outside of Paul's journey. Indeed, Paul's intention is to use the Schumann underground parking lot entry, which is situated before the incident site in this one-way street.

Therefore, the navigation system sticks to the current intentions (even when it might no longer be feasible) in a first phase. In contrast, navigation systems that have their journey cross the affected road segment start adapting their plans immediately. They will delay their arrival in the Wetstraat or select an alternative route that avoids the incident. Possibly, these affected travelers overreact, at which point unused capacity around the incident site will become visible to the predicting IM, so that the capacity is likely to be reserved again.

This D4U traffic coordination technology is model-based, where the models support virtual execution of travel activities. Support for multimodality was recently introduced simply by adding the required models reflecting the services offered by public transport. However, the synchronization of these models (e.g., reflecting delays and rescheduling triggered by such delays) is still under development. This causes many users to specify the public transport option as a backup only. The incident triggers such users affected by the incident to consider public transport, which is likely to reduce its impact.

Note how this system uses its intricate knowledge of the traffic infrastructure and user intentions to avoid that "misery spreads to travelers that have no business with it." Indeed, after this first phase, inflow of travelers that would block others when queued at the incident site has been reduced, and actually more capacity has become available to travelers who need to use the first sections of the Wetstraat only. Unfortunately for Paul, too many cars that cannot avoid the choking point in the Wetstraat are already queued up in the tunnel leading to the Avenue de Cortenbergh.

Therefore, his navigation system redirects him to another tunnel to enter Brussels. Because the weather forecast predicts dry weather with a high probability, the navigation system explores switching to a different parking lot based on Paul's profile, stating that he is willing to walk up to one kilometer. However, this proved to be unnecessary when the effect of the restraining of cars that need to pass the incident site became visible in the forecast, and as expected, revealed that higher capacity became available in the first sections of the Wetstraat. In the end, Paul drove his car to the originally intended parking lot along a slightly longer route.

MIRROR THE WORLD-OF-INTEREST (WOI)

To apply our first design principle (see Chapter 2), the main mechanism is to mirror the WOI in software. This first D4U principle can be paraphrased as "do not rely on expectations/assumptions that might be proven wrong." Note that the principle itself is broader and goes beyond WOI mirroring.

In transport and navigation applications, maps constitute a very old application of this principle and its implementation mechanism. On condition that a map is accurate, integration conflicts and other issues involving such a map cannot be attributed to the map. For instance, when a routing algorithm steers a truck under a bridge with insufficient clearance, as indicated correctly on the map, the issue needs to be addressed in the algorithm (i.e., it must be enhanced or replaced). Changing the clearance indicated on the map (but not in reality) will not solve this issue.

Indeed, inaccurate WOI reflection is the only situation in which the map needs changing to solve a conflict or address an issue. This can be a correction of an error. It can be an update (e.g., nautical maps of coastal waters may be updated biannually). Furthermore, it can be missing information (e.g., indication of one-way streets).

Missing information typically can be added without invalidating the existing information or breaking other applications that are using the maps but do not need the missing information. Indeed, changes to maps only involve restoring and/or enhancing its WOI reflection, irrespective of the nature of the issue that needs addressing.

Note that a system or organization that maintains such an accurate WOI reflection will comply – to a greater extent – with the first D4U principle. Indeed, it will adapt/update before an issue or conflict can arise. An example is the IM of the road segment in the above scenario. It is cooperating with the onboard safety systems in the crashed cars and the emergency services to keep its WOI reflection up-to-date, accurate, and complete.

Remark that maps, and WOI mirror images, only provide part of a solution. For example, in navigation systems, maps need to be complemented by a global positioning system (GPS) and a route-generating system. The GPS is compliant with the first D4U principle. However, the route generation will be inherently unable to comply with the principle in full. Recall that our research investigated how much of an overall application can be designed for the unexpected and which parts still need to be developed using more conventional approaches.

MOVE AS MUCH AS POSSIBLE INTO THIS WOI MIRROR IMAGE

In the applicability range of D4U, the WOI resides in the real world. And reality always is coherent and consistent (but not necessarily as we wish it to be). Whenever some WOI mirroring software contributes to achieving our objectives, that part of the system software inherits this coherency and consistency. In a way, the Creator of the real world is contributing to our design. Obviously, it is ill advised to ignore this Creator's contributions when available. Indeed, our research has progressed whenever it moved functionality – even partially – into the WOI mirroring software.

Accordingly, design for the unexpected aims to move as much functionality as possible into a reflection of the WOI. Research prototypes have been able to do so for the following:

- *Resources*. This is historically the oldest and most common category. For example, road maps are reflecting a road infrastructure. We distinguish two subcategories: the map legend (mirroring resource types) and the map itself (mirroring resource instances). State-of-the-art information and communication technology (ICT) allows to develop sophisticated mirror images, reflecting both properties (e.g., number of lanes) and state of resources (e.g., availability of a parking space). The mirror image may cover the present (track), the past (trace), and even the future (e.g., reservations).
- *Activities*. Commuting, holiday traveling, shopping, etc. can be mirrored in a manner similar to resources. Likewise, it is possible to distinguish activity types (how to complete a journey) from activity instances (executing such a journey). Mirror images utilizing state-of-the-art ICT allow a virtual execution of such activities on the mirror image of the resources to assess feasibility and expected performance. It also allows for track-and-trace implementations. Taking the future into account is covered as well.
- *Mental states and commitments*. Everything that exists in the WOI is candidate for inclusion in the mirror image. Caution must be given to restricting the development to mirroring what exists. When reflecting mental states and commitments, accuracy is the key concern. The expressiveness of the representation must cope with the corresponding reality such that no interpretation or interfering occurs. In mobility, intended journeys and train reservations are examples of what mirror images may reflect. Virtual execution of mental states – in particular, intentions – constitutes the starting point from which future states of resources and activities may be predicted (see Sections "Bioinspired Coordination and Control in Holonic Execution Systems" in Chapter 5 and "The DMAS Architectural Pattern" in Chapter 6).
- *Policies and decision-making mechanisms*. Within the WOI, there are a number of decision-making elements: humans, the laws of nature, simple rules (e.g., first come, first served), complicated policies (e.g., a finite-capacity scheduler), etc. Such mechanisms can be mirrored as well. For humans, some machine learning may be indicated. Simple rules can be their own model. The laws of nature and complicated policies may require approximations if this mirror image needs to virtually execute activities (a lot) faster than reality.

While developing the WOI mirroring software, the following concerns need addressing:

- *Completeness*. As a general rule, the mirror image should be complete such that its users (i.e., other software) always remain within this image while exploring and/or virtually navigating and executing activities. For instance, the WOI mirror image covers space and time from minus infinity until plus infinity. Note that this image may and will be very coarse at these extremities, reflecting that they are mostly irrelevant for the application. Completeness prevents users from "crashing when they fall off the world" because the WOI mirror extends to infinity. Completeness provides the necessary starting points for future expansion and enhancement of this mirror image without breaking the preexisting users. Indeed, completeness confines future adaptations to refinements of the current WOI image. Completeness implies that even highly undesirable and unlikely states are mirrored, which enables ICT support for recovery. Here again, mirror images may and will be coarse (i.e., only a small effort is needed).

- *Single source of truth (SSOT).* In the WOI mirroring software, everything in the WOI has a single counterpart. For instance, a parking space cannot be mirrored twice and create the illusion that it can be occupied by two cars at the same time. Our research results enabling to predict the unexpected are instrumental in reconciling this SSOT requirement with the need to collect and process information from multiple sources to deliver services and functionality.
- *Updating and accuracy.* A proper and correct implementation of a WOI mirroring software needs updating whenever the corresponding reality changes. When the composition of a team changes, the mirror image needs to adapt. In particular, mental states may change often, which implies frequent updating of the image. Moreover, the mirroring needs to be accurate. For instance, a mental state is nothing more than a mental state (e.g., the intention to stop smoking). Likewise, decision mechanisms are only mechanisms (e.g., first-come, first-served). On their own, they have no authority over the reality in the WOI. The WOI image must not succumb to wishful thinking. Adding assumptions to the reality that is mirrored is considered an inaccuracy.

MINIMIZE INERTIA OF DESIGN CHOICES

Application of the first D4U principle (cf. Chapter 3) results in an infrastructure mirroring the WOI. However, it is impossible to implement a system answering all user requirements in such an infrastructure. At some point, option-excluding design choices have to be made, which potentially may cause serious issues in the future.

Consequently, a second D4U principle requires that the inertia of the option-excluding design choices must be kept low. The corresponding generic mechanism to keep this inertia low consists of *explicit and mandatory resource allocation*. This means that every decision-making component has to obtain rights over the resources it requires before it is able to perform and offer its services. It also means that resources can be deallocated. Again, the D4U principle itself is broader and goes beyond the mechanism.

This resource allocation mechanism provides an upper bound on the damage that can be inflicted by design choices causing issues and conflicts. Indeed, the worst case consists of deallocating the resources from a nonperforming component and allocating them to a suitable one. If needed, such a suitable component has to be developed. This deallocation can be preemptive if indicated.

Note that resource allocation awareness will link the quality of decision-making components to their ability to minimize their requirements for rights over resources. Note that such minimizing is an application of the first D4U principle, which improves the ability to coexist with other decision-making components.

A SAMPLE SCENARIO (CONTINUED)

In the sample scenario, the road infrastructure elements are mirrored in executable software models. These elements are the resources. On these resource models, traveling activities are virtually executed beforehand and much faster than in reality. As resources – the road segments – only have a single counterpart (SSOT), each resource will be informed about all future visits from traveling activities as they virtually execute their journey. And resources will be informed repeatedly to account for any changes that occur in the mental state of the travelers (i.e., a change in mind) as well as on the roads themselves (e.g., a car crash).

Resource models use (dynamic network loading) models to compute how congestion is likely to propagate (backwards). The results are used to update travel time estimates for the road segments, which are then used by the travel activity models to predict their future journeys in time and space.

Models of the decision making by travelers are evaluated – during the above virtual travel execution – to determine the mental state of a traveler (i.e., to select the journey to execute virtually and inform the affected resources). Other decision-making models are evaluated to steer the search for candidate journeys through virtual execution; these models determine which part of a combinatorial solutions space will be investigated.

In other words, the D4U intelligent traffic and transportation system (ITTS) addresses most of the challenges through executable models of its WOI. The parts of the solution that fail to comply with the first D4U principle, that is, the decision-making elements, are considered to be part of the WOI. Executable models take care and handle those elements, delivering a proactive coordination to road travelers. These decision-making elements may and do use information about future/predicted resource loadings and activity routings.

Whenever the WOI changes, the repeated virtual execution of travel activities will update the information for the coordination. This virtual execution will evaluate the most recent model for its decision-making elements. In other words, it is a model-driven system that has minimal inertia concerning these models. Accommodating changes in resources, activities, mental states, or decision-making is automatic and effortless except for the requirement to have the necessary executable models.

In a way, the development of a D4U ITTS is only beginning because the decision-making elements – determining performance – still need to be provided. However, the D4U infrastructure can be deployed already and, on top of it, these decision-making elements can be put into operation with little effort as soon as they are available. Also, these elements can be replaced while the ITTS remains operational, can be personalized, etc. In the past, by focusing on the decision making first, we have been investing in disposable ICT. With D4U, a durable ICT platform and infrastructure can be developed, which will boost the development of high-performance decision-making solutions as their deployment time and effort decrease.

ABBREVIATIONS

D4U	Design for the unexpected
ICT	Information and communications technology
IM	Intelligent module
ITTS	Intelligent traffic and transportation system
SSOT	Single source of truth
WOI	World of interest

CHAPTER TWO

On the Design of Complex Systems

Paul Valckenaers*, Hendrik Van Brussel**
*Faculty of Engineering Technology, KU Leuven
**Faculty of Engineering Science, KU Leuven

This chapter establishes a context of our design for the unexpected (D4U). The various levels of system complexity are defined, and the superiority of structural design over functional design for D4U is highlighted.

ON SIMPLE, COMPLICATED, COMPLEX, AND COMPLEX-ADAPTIVE SYSTEMS

Systems, or problems, can be simple, complicated, or complex. The distinction has to do with the number of components and their interactions. A car is complex relative to a bicycle, but very simple relative to a manufacturing plant or an economy. One could say that a bicycle, containing a hundred components, is a simple system, and a car, with some ten thousand components, a complicated system. A manufacturing plant, with many more components, would also be a complicated system, but it is more, it is also a complex system. Why?

The behavior of both simple and complicated systems is well predictable. If one follows the assembly rules for a bicycle or a car, the behavior of the assembled system is predictable by knowing the starting conditions, because the interconnections between the system components are well-defined and fixed, while the component interactions remain simple and predictable.

In a complex system, the same starting conditions can produce different outcomes, depending on the nature and the sequence of the interactions between the system components. We are unable to understand the system by observing its constituents in isolation. At that point, we say that there is some "emergent behavior" or even self-organization and we declare the system "complex." The whole is greater than the sum of the parts. Self-organization goes beyond emergence; it is the ability of a system to spontaneously arrange its components or elements in a purposeful manner, under appropriate conditions but without the help of an external agency.

Design for the Unexpected. http://dx.doi.org/10.1016/B978-0-12-803662-4.00002-3

For example, building a highway is complicated, but managing urban traffic congestion is complex. Likewise, building a state-of-the-art air traffic control center is a complicated challenge in executing the project, while directing air traffic is complex, involving real-time problem solving. In the same way is a manufacturing system a complex system, hence showing emergent behavior. This book provides a framework on how to control suchlike systems.

While a complex system consists of a large number of components (often called "agents") that interact, the term *complex-adaptive system* (CAS) refers to a complex system in which the components (called "holons" in this book) not only interact but also adapt and/or learn. Self-similarity is often observed in a CAS. *Adaptivity* gives a complex system *robustness* (resilience) against disturbances and *autonomic behavior* (homeostasis).

An example of such adaptive behavior is found in the human immune system, which allows for vaccination to be an effective measure in preventive healthcare. But, it also works the other way around; inadequate therapy adherence has induced bacteria to become resistant against many/most of the known antibiotics. Typically, this adaptive behavior represents a challenge. For instance, commuters will react differently when receiving information from an intelligent traffic system (ITS) based on their past experience (of congestions), and they may change collectively (e.g., triggered by a weather forecast), often rendering the ITS less effective.

Optimality is an important issue when designing complicated or complex systems. F.W. Taylor, the father of scientific management, claimed that a complex system/organization would be optimal when each of its components were optimized separately. This *Taylorian* view stands perpendicular to the present generally accepted view that optimality can only be achieved if the complicated/complex system is considered in its entirety and subject to optimization as a whole. For instance, Goldratt's theory of constraints identifies the utilization rate of a system's bottleneck as the determinant for system optimality (i.e., throughput in nonstop production) (Goldratt, 1984). However, the location of this bottleneck is system and situation dependent.

Mechatronics is the discipline advocating a concurrent or simultaneous engineering approach to the design of "optimal" complicated/complex artifacts, in contrast with the traditional sequential engineering approach, where the different aspects/behaviors are dealt with in a sequential order. This book does not deal with designing mechatronic systems as such. It does however deal with designing, or at least controlling, systems of mechatronic systems or of what are called here *mechatronic societies* (MSs).

A mechatronic system has a self-similar (fractal) nature when looked at on a component, machine, and machine-system level but is in itself mostly not a complex system, rather a complicated system. A system of mechatronic systems, called "mechatronic society," is a complex system, often adaptive.

Typical mechatronic societies are manufacturing systems, factories, traffic systems, etc., consisting of a set of rather autonomous, mechatronic systems, called agents or holons, interacting with each other and with the outside world, in order to achieve an overall system goal. Such systems will be called "holonic systems" in the remainder of this book. They behave optimally, not in the mathematical sense of the term but in a more useful way, as explained later. MSs are complex-adaptive systems, in that they adapt their behavior as a function of changes in their structure (e.g., machine breakdown) or in the environment (e.g., road block). They exhibit emergent and self-organizing behavior, as will become clear later on.

ON THE DESIGN OF COMPLEX-ADAPTIVE SYSTEMS

Both design teams and individual designers face limitations concerning the complexity of the artifacts that they may develop. When building complex systems, designers develop subsystems that are integrated into a larger system. The resulting system exhibits an emergent behavior that was never explicitly planned or conceived by these designers simply because its complexity exceeds their mental capabilities.

Many industrial design teams, designing for instance automobiles, remain largely in control of what their artifacts will be. In contrast, many of the most valuable artifacts in a modern human society, especially infrastructures, simply are too complex to be conceived explicitly by humans. They emerge by the combination and integration of simpler systems, which form their constituents. Unfortunately, the resulting emergent behavior is too often characterized by poor performance, missed opportunities, and the inability to serve smaller user communities.

In the next chapter, the fundamental nature of the earlier issue is addressed. What causes the difficulties occurring when smaller systems are combined and integrated into a larger and more complex system? What can be done to remedy the integration/emergence problems, which are observed in reality? This research builds on the ideas of Herbert Simon (Simon, 1990) as well as insights discussed in Waldrop (1993). Waldrop gives a readable and concise overview of relevant developments in the domain of complex-adaptive systems.

Simon emphasizes the issue of the fundamental difference between analytical (i.e., observing) sciences and synthesizing (i.e., engineering, designing, creating, etc.) sciences. Every artifact design is, to a significant extent, arbitrary; there are numerous ways in which a design problem can be solved "correctly." In contrast, the different manners in which the laws of physics can be described only differ in very superficial ways (i.e., the symbols that are used). In making this observation, Simon touches the core issue: a scientific understanding of synthesis/design is lacking. Accordingly, Chapter 3 aims at this "unavoidability" identified by Simon as a key property of a scientific understanding for synthesis/design.

Furthermore, the next chapter complements more classical work on design (Suh, 1997) aimed at design tasks where the design team remains largely in control and typically designs for a short time horizon. Such work typically builds on the decomposition of functional requirements (FRs) and top–down design of solutions. In contrast, Chapter 3 aims at the design of more long-lived artifacts that are part of an emerging overall system. Such artifacts are typically considered as part of the existing technology base in the shorter-term design situation. The work in this book therefore has a closer relationship to (some approaches within) object-oriented design, which is more targeted at problem solving while facing uncertainty and complexity that defies our mental capabilities. These two related topics are discussed later in this chapter.

TOP–DOWN FUNCTIONAL DESIGN AND DEVELOPMENT

In the early days, software engineering embraced the predominant development methodology from older engineering disciplines: top–down decomposition and functional design. Axiomatic design (Suh, 1997) is a representative example of this functional design approach (of artifacts rather than of software), among which it is more formally elaborated than most. This method starts from the FRs defined by the users. These FRs can be satisfied by manipulating some design parameters (DPs). Axiomatic design is based on two basic axioms: (i) the independence axiom and (ii) the information axiom. The independence axiom requires the FRs to be independent. A design is independent if each FR is controlled by only one DP. The information axiom states that the best design is the simplest design that still satisfies all the FRs, where these FRs are considered to be independent (no overlap); if necessary, user requirements must be reformulated. Next,

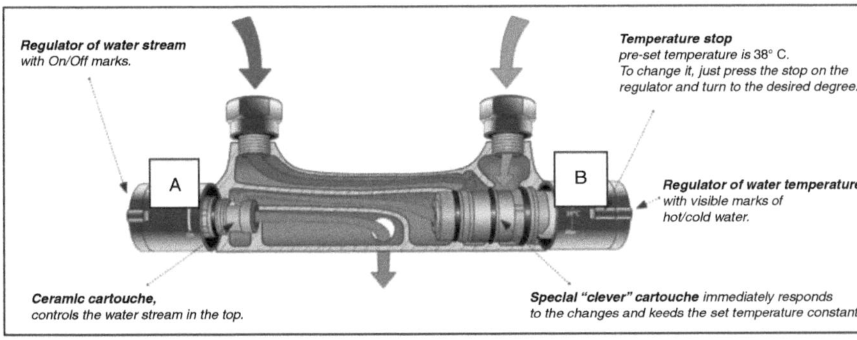

Figure 2.1 ***Thermostatic Faucet Designed According to the Independence Axiom. (Source: RAVAK a.s.).***

the top-level design answers these requirements. A thermostatic faucet is a good example of an independent design. The two FRs (water temperature and flow rate) are controlled by two independent DPs (two independently adjustable regulators A and B on Figure 2.1).

If the top-level design cannot be implemented in a straightforward manner, each of its components becomes a set of requirements and the same design process is repeated at the component level. If necessary, a second-level design is again decomposed, and this is repeated as long as required to obtain components that can readily be implemented. Consider the design of a deep freezer as an example. The user requirements comprise cooling performance, storage volume, energy efficiency, and accessibility. Based on a specific set of requirements, a designer may develop a chest-type deep freezer as his/her top-level design. The respective parts of the chest become the requirements for the next-level design activities. The door design will be different for a chest type from that of a cabinet type, for example, from the viewpoint of energy efficiency.

This proven development methodology numerous times has shown to be effective. In particular, the focus on functionality generates highly efficient systems: it minimizes the complexity of the final solution as well as the effort, time, and resources needed to answer the given requirements. This property still makes this development methodology the best alternative when (i) answering the initially given top-level requirements is what matters, and/or when (ii) this efficiency is a truly dominating concern. Unfortunately for IT developments, these two conditions hardly ever hold. User requirements are notoriously unstable, and computer programs can be much larger and/or slower than theoretically necessary. The latter is especially true when allocating ample computer memory and processing power reduces software development and debugging or maintenance time and efforts.

The attractive property of top–down functional development – optimized efficiency – is also its weakness: artifacts under stress, caused by such optimization, are fragile. The typical result is exceedingly vulnerable to changes in the user requirements. Consider our chest-type deep freezer again. After using the freezer for a short time, the users realize that their initial accessibility requirements were too weak (they fail to remember what is stored at the bottom). Repeating the design process now yields a cabinet-style deep freezer with drawers. Second-level and higher-level designs cannot be reused and become virtually useless; remember that these levels rely on the given context to yield an efficient answer to their given requirements. The elements that survive the change in requirements are bottom-up technological developments (e.g., thermally isolating materials) and generic design know-how (e.g., containers with the opening at the top: chest in the first design, drawers in the second).

Furthermore, top–down functional design forces the designers to make important choices early in the process when they have minimal knowledge (e.g., select a chest or cabinet design in the deep freezer example). In software design, developers often have to learn while they develop. Undoing early unfortunate design choices can be extremely difficult because it invalidates much of the subsequent developments when following this conventional approach.

In conclusion, top–down functional design is suitable when (i) the user requirements are stable, (ii) the designers already have much experience in the system they develop, and (iii) efficiency is important in the sense that the design may not be more complex and consume more resources than necessary. When building a large bridge, such conditions are likely to be fulfilled: (i) the way people use bridges has remained unchanged for ages; (ii) the main design task consists of adapting proven designs to the particular circumstances in which this bridge has to be built; (iii) efficiency is crucial because, unlike software, a bridge cannot be twice the necessary size. In other words, top–down functional design remains the methodology of choice for developments that need to become operational as fast as possible, need to be efficient, and can recover their costs by answering the initial requirements only. Such developments typically produce macroscopic mechatronic systems or mechatronic societies (e.g., buildings, ships, machines, chemical plants).

However, the preconditions for justifying a top–down functional design methodology to software development are not fulfilled. Indeed, development costs for reproduction of a solution become increasingly less important for high-tech products, systems, and infrastructures, whereas user requirements become less stable, and learning during the development becomes more important. The answer to this issue in the domain of software engineering is the subject of the next paragraph.

OBJECT-ORIENTED DESIGN AND DEVELOPMENT

Object-oriented technology is widely recognized to be a most significant development in IT technology in the last few decades. Object-oriented programming languages, such as C++ and Java, have become the dominant programming tools for leading professional software developers. An object-oriented design methodology complements these object-oriented programming languages. It is object-oriented design that mirrors the deeper insights on how large, complex IT systems can be successfully developed; especially while user requirements keep changing.

Already in the early 1980s, Michael Jackson recognized user requirements to be among the highly unstable parts in a software development activity. Jackson (Jackson, 1995) also recognizes other weaknesses in top–down design (e.g., being forced to take crucial decisions early, the absence of a single hierarchical decomposition in the real world) and proposes his Jackson system development (JSD) methodology as an alternative.

Jackson recognizes that the world of interest, for the software under development, is much more stable over time. Jackson's methodology consisted of reflecting the entities of interest and their relationships in software first. This comprises, for instance, the information related to the employees of a company (e.g., name, contract, working hours, and working place). Next, measures to keep the information in the computer system synchronized with reality are implemented (e.g., when a person gets promoted, the appropriate data fields are updated). Finally, the functionality needed to answer the user requirements is implemented on top of this reflection of the world of interest (e.g., management reports on the monthly expenses for wages). Jackson's methodology was applied for the development of COBOL applications in those days.

Jackson's ideas constitute the core of today's object-oriented design methodologies. The work by Cook (Cook and Daniels, 1994) constitutes a representative example of such methodologies. In an object-oriented design, the developer implements an *essential model* reflecting the world of interest for the problem that needs to be solved. Such an essential model describes (i) the possible states the world-of-interest can be in, (ii) which events cause which state transitions, and (iii) the possible sequences in which events can occur.

In the implementation, the software keeps the essential model synchronized with reality. Functions are implemented on top of this essential model.

The core ideas of object-oriented design had difficulties disseminating themselves through the world of software designers. First of all, the core message often disappears in the information about formalisms and notations (e.g., UML diagrams), and the information related to object-oriented programming. Second, object-oriented design is not suitable for all software development activities (e.g., the design of quick-and-dirty solutions for which a low developmental effort is more important than the solution becoming useless in sometime in the future).

Nonetheless, object-oriented thinking is well established among current IT professionals as put to evidence by Nelson et al. (2002), who discusses a compact course to teach older expert functional software developers to adopt object-oriented design. The key element of this course is to make these experienced developers themselves discover that user requirements (functions) constitute an unstable element in their world and thereby make them discover the essential model as the stable part.

The essential model – using the wording from Cook and Daniels – has become a part of widely recognized object-oriented software engineering methodologies, commonly referred to as (variations of) the unified process; note that the terminology may differ, depending on its source. A key feature in this unified process is its architecture-centric nature. Likewise, design for the unexpected (D4U) focuses on an architecture in which essential models – ideally in executable implementations – have a most prominent position indeed.

COLLECTIVE DECISION MAKING AND ARCHITECTURE-CENTRIC DESIGN

Complex systems usually are the result of a collective effort. In such community efforts, the argumentation of collective decisions often relies on abductive reasoning to justify and create the perception that a given group consensus is "the" solution. However, such reasoning can be classified as (cf. appendix III) (i) *deductive reasoning*: conclusion guaranteed; (ii) *inductive reasoning*: conclusion merely likely; and (iii) *abductive reasoning*: taking your best shot.

In fact, abductive reasoning is only a heuristic where the outcome – this group/community consensus – receives too much respect and has unjustified authority in today's practice. The abductive reasoning, leading to this consensus, is considered *de facto* proof of correctness and unavoidability, where abduction only delivers a possible option out of many. Specifically, collective decision making commonly uses abduction to justify a good enough solution respecting the comfort zones and vested interests of the controlling parties and stakeholders.

As will be discussed in the next chapter, design for the unexpected addresses problems by first elaborating intermediate solutions – designed for the unexpected – from which final solutions – solving the problems – are elaborated. To achieve design for the unexpected as the result of a collective effort, a collective awareness of this divide and its purpose/benefits is required. Architecture-centric development approaches address such concerns. The software engineering community enjoys such a collective awareness of the contributions by suitable architectures. Other communities may still need to discover this.

SUMMARY AND REMARKS

Without attempting to be comprehensive, this chapter presents a context in which D4U is situated. It distinguishes simple, complicated, complex, and complex-adaptive systems where the latter have become, and continue to be, more and more important in our society.

Next, design by means of top–down functional decomposition is discussed, highlighting axiomatic design. It is a valid approach when addressing shorter-term problems with stable requirements. Complementing this approach, D4U provides, creates, enlarges, and improves what is considered the technology base by functional design. On the whole, D4U facilitates the application of axiomatic design.

There exist related/similar approaches. For instance, role-based decomposition is promoted in *the autonomous agents and multiagent systems* (AAMAS) community. Such approaches share with functional decomposition that this role decomposition is human-invented, which prevents offering scalability or integrate-ability guarantees.

Structural decomposition, found in essential models within object-oriented design, is addressed subsequently. This matches D4U well. However, in practice the contribution and importance of these essential models is ill-appreciated. The fact that there is no single widely used terminology for the concept of essential models reflects this. Essential models mirror a reality that is coherent and consistent. When designed along D4U principles, they scale and integrate well.

Furthermore, D4U shares the architecture-centric nature of object-oriented software engineering methodologies (see Chapters 5 and 6). However, software objects are passive, requiring method calls to perform actions. To effectively mirror reality, active communicating computing processes are needed, often called agents. Therefore, the software implementation technology of choice for D4U will be an "actor language" rather than an object-oriented language. Indeed, interaction is much more powerful and expressive than computation (Wegner, 1997).

ABBREVIATIONS

AAMAS	Autonomous agents and multiagent systems
CAS	Complex adaptive system
DP	Design parameter
FR	Functional requirement
IT	Information technology
ITS	Intelligent traffic system
JSD	Jackson system development
MS	Mechatronic society
UML	Unified modelling language

REFERENCES

Cook, S., Daniels, J., 1994. Designing Object Systems. Prentice-Hall, London.

Goldratt, E., 1984. The Goal. North River Press, Great Barrington, MA.

Jackson, M., 1995. Software Requirements and Specifications. Addison-Wesley, Amsterdam.

Nam Suh, P., 1997. Design of systems. CIRP Ann. 46 (1), 75–80.

Nelson, H.J., Armstrong, D., Ghods, M., 2002. Old dogs and new tricks: a compact course teaches the expert procedural programmer to think OO. Commun. ACM 45 (10), 132–137.

Simon, H.A., 1990. The Sciences of the Artificial. MIT Press, Cambridge, MA.

Waldrop, M.M., 1993. Complexity: The Emerging Science at the Edge of Order and Chaos. Simon and Schuster, New York, 380 pages.

Wegner, P., 1997. Why Interaction is more Powerful than Algorithms. Communications of the ACM 40 (5), 80–91.

CHAPTER THREE

Design for the Unexpected

Paul Valckenaers*, Hendrik Van Brussel**
*Faculty of Engineering Technology, KU Leuven
**Faculty of Engineering Science, KU Leuven

This chapter presents a theoretical analysis/model of how designers elaborate systems that can handle unpredictable user requirements and integration requirements. No advanced know-how is required to understand this analysis; elementary set theory suffices (Cf. http://en.wikipedia.org/wiki/Set_theory – basic concepts and notation.). The analysis provides insights allowing designers to enhance their design and development process when they appreciate precisely when, where, and how design choices affect the ability to cope with the unexpected.

PROBLEMS AND SOLUTIONS

This section formally addresses what constitutes a problem and its solution(s). The purpose of the formal approach is to present ideas more precisely. The formal approach does not produce any calculus on problems and solutions, nor does it claim completeness in a philosophical sense. The formalism mainly serves to avoid ambiguity and to delineate the ideas more sharply than would be possible in natural language.

Problems

A problem P is defined as follows:

> A problem P is a constraint on the state space $\mathbf{U}$ of the universe U, defining a set $\mathbf{P} = \{\mathbf{u} \in \mathbf{U} \mid \mathbf{u} \text{ satisfies } P\} \subseteq \mathbf{U}$.

When U is the world in which we live, $\mathbf{U}$ is an infinite state space. Every state $\mathbf{u} \in \mathbf{U}$ has a time coordinate $\mathbf{t}_{\mathbf{u}} \in \Re$. By definition, *reachable states at a given time coordinate* are states that either have been or can become the actual state of the universe at this given time coordinate. This universe is subject to the laws of nature (constraints), which limit the number of states that are

Design for the Unexpected. http://dx.doi.org/10.1016/B978-0-12-803662-4.00003-5

reachable. These laws of physics imply that there is exactly one reachable state $\mathbf{u}$ for every $\mathbf{t}_{\mathbf{u}} \leq \mathbf{t}_{\mathbf{now}}$.

The universe *U* follows a trajectory through its state space as time progresses. This trajectory is defined for states up to $\mathbf{t}_{\mathbf{now}}$. It consists of the states in which the universe has been in the past. The future trajectory is only partially defined. This future trajectory is constrained by physical laws, possibly including stochastic aspects, and is affected by the actions of the human and other entities in this world. These actions affect the choice of the successor states of the current state corresponding to $\mathbf{t}_{\mathbf{now}}$. Normally, any significant impact on the trajectory requires sustained action during a substantial amount of time. A problem *P* is solvable by an agent (human or otherwise) if the agent is able to make this trajectory stay within the given subset **P**.

Remark 1

In our world, there exist two major types of problems (Wegner, 1997). First, there are the *one-shot* problems, which require the state of the universe to comply once with a constraint at some point in time. The problem specification does not care about the states before and after. An example is to deliver in time some quantity of goods of sufficient quality. Such problems often are agreements between humans to coordinate their interactions. The second type of problems consists of *going concerns*. Going concerns require that the trajectory of the universe complies with requirements that span a complete time window, typically starting from $\mathbf{t}_{\mathbf{now}}$. For instance, a coordination and control system must keep its underlying system in a safe state (no casualties etc.). Note that there exist many problem-solving technologies that cannot handle going concerns (e.g., database query engines and most optimization software). Inherently, most real-life problems are going concerns, where one-shot problems often are artificial problems. Accordingly, coordination and control technology normally addresses going concerns, possibly using one-shot problem solvers as subsystems.

Remark 2

The above defines a "basic" problem as a constraint with which a solution must comply. In reality, there also exist optimization problems. These optimization problems can be modeled as a set of tuples, where each tuple consists of a basic problem and a valuation, where the overall problem is to optimize this valuation. This valuation defines the value/benefit/… of solutions to the corresponding basic problem. Such a valuation can be single-valued and fully ordered, but it can also reflect multicriteria optimization problems. Since the remainder of the discussion does not require this extension to optimization problems, it is not elaborated further.

Solutions

A solution S to a basic problem P is defined as follows:

> A solution S of a problem P is a constraint on the state space $\mathbf{U}$ of the universe U defining a set $\mathbf{S} = \{\mathbf{u} \in \mathbf{U} \mid \mathbf{u}$ satisfies $S\}$, where $\mathbf{S} \subseteq \mathbf{P} \subseteq \mathbf{U}$ and $\forall\, \mathbf{t} \in \Re, \exists\, \mathbf{s} \in \mathbf{S}: \mathbf{t} =$ timeCoordinateOf($\mathbf{s}$).

Agents (human or otherwise, intentionally or unintentionally) solve a given problem P when they confine, through their actions, the trajectory of the universe U to the corresponding subset $\mathbf{P}$. Therefore, their actions – combined with the laws of the universe – correspond to constraining the state of the universe to a subset $\mathbf{S}$ of $\mathbf{P}$. $\mathbf{S}$ cannot be empty; it will always have at least one state for every time coordinate. If S fails to comply with this condition, the agents failed to solve the problem.

Typically, $\mathbf{S}$ and $\mathbf{P}$ will differ significantly concerning the states with a time coordinate that is smaller than or equal to $\mathbf{t}_{\mathbf{now}}$. The problem P is only concerned with what is needed, useful, and so on. Therefore, it allows as many states as possible as long as the choice among them does not matter. In contrast, the solution S is embedded in the universe, which allows only a single state for every time coordinate in the past (including the present). Moreover, S will reflect that are severe limitations on what states can be reached in the immediate future from the current state (e.g., it takes time to travel from A to B, to prepare a meal, or to build a house).

In other words, $\mathbf{S}$ will be significantly smaller than $\mathbf{P}$, especially concerning states close to the present time or in the past. Therefore, problem-solving agents unavoidably have to make choices whenever deadlines approach; they have to select a single state (per time coordinate) from all states allowed by the problem (for the time coordinate). An example of the introduction of constraints can be observed in the design of a railway system to solve transportation problems: when the moment of actual usage approaches, the designers have to make more and more choices. For instance, they must select a specific value for the track width.

In real life, a problem-solving activity consists of a sequence of actions over time. Using the above definition, such sequence of actions corresponds to a sequence of solutions S_1, S_2, … S_{end} that solve P, where $\mathbf{S}_{\mathbf{end}} \subset \cdots \subset$

$\mathbf{S}_2 \subset \mathbf{S}_1 \subset \mathbf{P}$. This reflects that *the agents make more and more choices as time progresses in order to solve the problem.* This introduction of constraints by the solution – as the agents make these choices – is a key issue. Indeed, when confronted with new requirements (unexpected ones), S_x may experience a conflict while P has no issue accommodating these requirements.

EMERGENT SOLUTIONS AND INTEGRATION PROBLEMS

When solving real-life problems, agents – human or otherwise – start from existing subsolutions, technology, and infrastructure. Adding to these existing parts, individual agents and agent teams provide new parts for an overall solution. These parts, existing or new, are brought together, integrated as far as possible, and an overall solution emerges. In other words, such emergent solutions involve integration where the integrate-ability of building blocks determines what may emerge, what performance will be achievable.

In this situation, the individual agents or teams face a high level of uncertainty about the other parts with which their part needs to cooperate. Typically, the agents contribute to the solution of many problems over time (e.g., a section of a transportation infrastructure is used in solving numerous individual transportation problems). Moreover, an agent's contribution is used to solve problems unknown at the time this contribution is created, and it must be combined with contributions from other agents. Many of these other contributions are developed independently such that the individual agent has limited opportunities to coordinate its contribution with the others. Some of these other contributions only emerge after the creation of the contribution of such an individual agent. In other words, integration problems confront existing subsolutions with requirements that were unknown at their design time; unsurprisingly, these requirements often will be "unexpected."

Formally, agent **x** solves problem P through solution S_p. The other agents solve problem Q through solution S_q. This results in the state sets $\mathbf{S}_\mathbf{p} \subseteq \mathbf{P} \subseteq \mathbf{U}$ and $\mathbf{S}_\mathbf{q} \subseteq \mathbf{Q} \subseteq \mathbf{U}$. For instance, agent **x** has constructed the railway system in France to answer the need for transportation. The other agents implemented similar railway systems in the remainder of Europe. Subsequently, providing transportation all over Europe – that is, problem T – is to be solved by integration of the national railway systems — that is, integration of S_p and S_q into solution S_t. As discussed in the next chapter, such large complex systems can only be created and sustained through combining subsystems/subsolutions; construction from scratch is too expensive and time-consuming.

Formally, agents must solve problem T, where $\mathbf{T} \subseteq \mathbf{P}$ and $\mathbf{T} \subseteq \mathbf{Q}$. Practically, agents must integrate the existing subsolutions into their overall solution; formally, $\mathbf{T} \subseteq \mathbf{S}_p \cap \mathbf{S}_q$. In the example, society is unable to duplicate the effort of developing their national railway systems to provide an international one. Instead, it must reuse the existing system to create the international connections among the national systems and obtain a system that transports goods and passengers across the borders in Europe.

The main problem with the creation of such solutions is that the integration fails to deliver the desired performance. In the railway example, it is relatively easy to provide international transport at a reduced service level; goods and passengers need to change trains at the borders of Spain and Russia (where the distance between the rails differs from the rest of Europe). The subsolutions, which were developed independently, have made mutually incompatible design decisions, and these decisions have accumulated significant inertia (i.e., it has become costly to undo these decisions).

Formally, for any ambitious T:

$$\exists\, \mathbf{t} \in \Re, \forall\, \mathbf{s} \in \mathbf{S}_p \cap \mathbf{S}_q \cap \mathbf{T}: \mathbf{t} \neq \text{timeCoordinateOf}(\mathbf{s})$$

In practice, a solution for T cannot readily reuse the (partial) solutions offered by the individual agents without undoing a lot of design choices. Typically, society only receives a reduced level of service (i.e., solutions to easier problems), and will only gradually outgrow the old designs when technology progresses sufficiently to introduce a new solution from scratch (e.g., high-speed trains).

The key issue is the introduction of constraints, by the agents, that are absent in the corresponding problem and that may cause future integration problems. More precisely, it is the accumulated inertia – that is, the effort needed to undo such harmful design decisions – that constitutes the problem.

DESIGN PRINCIPLES – DESIGN FOR THE UNEXPECTED

It becomes clear that the design of subsystems for emergent solutions imposes its own requirements on a design activity. In particular, the problem-solving agent must design a solution S_p capable of surviving in an uncertain environment concerning its future.

Formally, such uncertain environment corresponds to $\wp$, a set of subsets of the state space $\mathbf{U}$. The actual future will offer an intersection of members of $\wp$ as the space that is available for S_p to contribute its part to the overall solution. Some members of $\wp$ correspond to the constraints imposed by the future problems for which solution S_p may be part of the overall emergent solution. Alternatively, members of $\wp$ correspond to the constraints imposed by other candidate subsolutions that may contribute to solving the bigger problem at hand. A designer of solution S_p does not know which members of $\wp$ will be present, and must try to avoid conflicts with any constraints that might be presented by members of $\wp$.

In this context, design decisions can introduce two types of constraints: stable and unstable. *Stable constraints* will be present in all conceivable future situations within the scope of S_p. For instance, an agent may assume that power supply will be 240 V/50 Hz or 110 V/60 Hz when designing an electrical appliance. Formally, no member of $\wp$ will be in conflict with the stable constraint. In contrast, *unstable constraints* represent conflicts with some members of $\wp$. Design decisions that introduce unstable constraints harm a solution's capability to contribute when solving future problems. For instance, a software designer may assume that two digits suffice to represent the year in a date field in a database design. In another case, the designer of a rocket inertial navigation program selected to minimize the memory and processing power requirements; this software only supported the acceleration range of the then-current rocket. When this software was reused, without the needed adaptations, to guide a more powerful successor, it caused the crash of this next-generation rocket during the first launch. It was a very expensive manner to detect a sub-solution integration conflict. A case with less serious consequences was an army's software system supporting only one data field to indicate size for clothing, which caused a problem when, sometime after this software was installed, women were allowed to become soldiers.

Based on this, design principles P1 and P2 emerge. In a way, they redefine and expand the "burden of proof" for a design decision to be considered justified. For example, in P1, the word *potentially* calls for a much more substantial justification than required in more conventional approaches.

P1. PROBLEM SOLVERS MUST AVOID INTRODUCING POTENTIALLY HARMFUL CONSTRAINTS

This design principle calls for design decisions introducing stable constraints first and as much as possible (i.e., certainly before introducing unstable constraints). These decisions are unlikely to cause future conflicts, as the introduction of stable constraints simply reflects the fact that these constraints are already present in the environment (cf. the universal use of 240V/50 Hz in Europe). A stable design decision preserves and reflects the scope of the problem domain.

For instance, using maps in navigation systems complies with P1. More generally, including essential models (see Chapter 2) in a solution complies with P1. In manufacturing systems, nonlinear process plans constitute a sample application of this first design principle. Nonlinear process plans indicate all possible manners in which a product can be manufactured rather than selecting a single sequence of processing steps.

The first principle discourages design decisions from introducing unstable constraints as they may create conflicts during future integration efforts. As discussed, it will be impossible to avoid introducing unstable constraints indefinitely, simply because t_{now} advances toward a deadline by which a problem needs solving. Therefore, the design principle requires unstable constraints to be introduced as little and as late as possible. It keeps options open as much as possible and as long as possible.

P2. PROBLEM SOLVERS MUST AVOID/REDUCE THE INERTIA BUILD-UP FOR POTENTIALLY HARMFUL CONSTRAINTS

When designers cannot avoid introducing unstable constraints, they may not use them to justify subsequent unstable design decisions, which would have introduced these potentially harmful constraints as well. Repeatedly making design choices introducing an unstable constraint will build up inertia: it will practically become impossible, take too much time and effort to undo this introduction when it reveals to be an unfortunate choice later. For instance, legacy systems are commonly associated with this issue.

Obviously, later design decisions have to be compatible with earlier design decisions, including unstable ones, in order to solve a problem. In addition, it is tempting to rely on all previous design choices as this will simplify the design and development task (i.e. there are less possibilities to consider); this is precisely what top-down functional decomposition in axiomatic design advocates. However, relying on earlier unstable design choices implies more design choices that need undoing when they reveal to be harmful (i.e. create a conflict).

Therefore, P2 basically states that P1 needs to be applied on the original situation: do not introduce potentially harmful constraints relative to the initial problem-solving situation. The fact that some earlier design choices already introduced potentially harmful constraints does not represent a permission to repeat their introduction.

CONCLUSION AND REMARKS

Summarizing, the novel principles for the designers are as follows: (i) designers must prefer stable design decisions and (ii) earlier unstable design decisions are no justification for later decisions imposing the same constraint(s). The first design principle avoids the introduction of new constraints. The second avoids the build-up of inertia for unstable constraints that were introduced earlier; every unstable design decision must be justifiable by itself.

The theoretical analysis/modeling and design principles in this chapter enable developers to judge more precisely when and how their design is protected against unstable elements in their environment. The theory reveals that "reflection of the world of interest" intrinsically is a source of stable constraints. Indeed, every element of the environment $\wp$ contains all the constraints originating from this relevant reality.

Chapters 5–7 discuss research results applying and translating the above to real-world application domains. Also, note that the above principles apply when designing lasting artefacts, such as infrastructures, and not for the short-time solutions for the immediate future.

ABBREVIATIONS

P Problem
S Solution
U Universe

REFERENCE

Wegner, P., 1997. Why interaction is more powerful than algorithms. Commun. ACM 40 (5), 80–91.

CHAPTER FOUR

Laws of the Artificial

Paul Valckenaers*, Hendrik Van Brussel**
*Faculty of Engineering Technology, KU Leuven
**Faculty of Engineering Science, KU Leuven

This chapter discusses four scientific laws or principles that apply to the artificial (i.e., to manmade systems). When bounded rationality determines what is possible or impossible, these principles are unavoidable and simply cannot be ignored by developers without consequences (i.e., failure, poor performance, or high costs). These laws concern flexible aggregation (holonic systems), critical user mass, decentralized steering, and collective proactiveness.

ON THE MEANING OF THE WORD *LAW*

Once upon a time in a small town, a civil engineer – employed by the town council – was instructed to elaborate a proposal for a new much-needed water reservoir to feed the town. The engineer designed a wonderful tower that was to be built on top of the highest hill in the vicinity, and she presented this to the town council.

The mayor appreciated the beauty of the design so much that he instructed the engineer to build the tower in the center of town. At which point the engineer replied, "That will not be possible because of Newton's *law* (of gravity)." Furious for not getting what he wanted, the mayor ordered the engineer to leave the room and wait outside.

After about an hour, the engineer was called in. Proudly, the mayor announced that they had solved the problem. They had checked all applicable legislation thoroughly and concluded that Newton's law was not applicable to their town.

The word *law* is used in at least two different ways: (i) manmade regulations and (ii) statements of facts (scientific law). Examples of the latter are the laws of nature, which are unavoidably true within their scope. Newton's laws on gravity and force–mass–acceleration are facts until you get outside their scope (e.g., move at speeds approaching the speed of light, go to subatomic dimensions). The first and second laws of thermodynamics constitute another example. This chapter uses the word *law* in the sense of laws of nature.

Design for the Unexpected. http://dx.doi.org/10.1016/B978-0-12-803662-4.00004-7

Recognition that there are laws of the artificial, resembling laws of nature, has been a major contribution from Nobel Prize winner Herbert Simon. Simon even wrote a book on this titled *The Sciences of the Artificial* (Simon, 1990). These laws of the artificial are statements of (unavoidable) facts in the manner the laws of nature or physics are.

If engineers fail to account for the law of gravity while designing a canal infrastructure, its performance will be poor while investments in pumping stations and recurring expenses to power these stations will be high. Likewise, ignoring the basics of complex manmade artifact design and development will result in disappointing performance and high costs.

This chapter discusses a (nonexhaustive) list of these laws of the artificial providing insights that are accounted for in the remainder of this book. The scope of these laws can be formulated "beyond a certain level of system complexity," which more conventional approaches fail to master. Indeed, this scope delineates those problems that benefit from and perhaps even require us to design for the unexpected.

AXIOMS

The laws of the artificial are derived from a number of observations of key properties of our world. They are assumed to be true (i.e., they are axioms, not lemmas).

Bounded rationality is a key element from which these laws are derived. Even a team composed of the most talented human individuals in the world, enjoying the best imaginable support and resources, has limited problem-solving and design capabilities. Moreover, enlarging such a team rapidly fails to improve its capabilities, i.e., when the increase in coordination and communication efforts absorbs more brain power than is added by the new team members.

A dynamic environment is another element from which laws of the artificial are derived. A dynamic environment implies that there is a time window in which problems need solving, designs have to be realized, and systems must be implemented. In combination with bounded rationality, this results in an upper bound on the number of information-processing steps leading to a design or solution to a problem. Above this upper bound, solutions arrive too late to contribute.

A competitive environment constitutes a third element. Whenever it is possible to improve, systems and designs that neglect to do this disappear or never appear at all. Again, possible-to-improve is to be seen in the finite time window allowed in a dynamic environment. Likewise, possible-to-improve is to consider the availability and recruitment of the necessary resources to execute the improvement. In particular, when bigger means better (i.e., more competitive, not necessarily more desirable), systems (of systems) will grow bigger.

Adaptation, the final element, relates to the fact that systems change over time. A solution may need to assume the world to be both complex (e.g., exhibit a butterfly effect) and nonrepeating (e.g., when past experience influences future behavior). Designers should realize that from a certain system size on, a perfectly predictable complicated system turns into a complex system, the behavior of which is becoming (partly) unpredictable, and for which new control approaches are needed.

LAW 1: HOLONIC SYSTEMS – FLEXIBLE HIERARCHIES

The term *holon* was introduced by Arthur Koestler in his book *"The Ghost in the Machine"* (Koestler, 1967). Two observations of how social and biological systems are organized motivated Koestler to propose the concepts of holons and holonic systems.

The Watchmakers' Parable (Simon, 1990)

Once upon a time, there were two watchmakers, named Hora and Tempus, who both made very fine watches. The phones in their workshops rang frequently, as new customers were constantly calling them. Hora prospered while Tempus became poorer and poorer. In the end, Tempus lost his shop. What was the reason behind this?

The watches consisted of about 1000 parts each. The watches that Tempus made were designed such that when he had to put down a partly assembled watch, for instance to answer the phone, it immediately fell into pieces and he had to start all over again, reassembling the basic elements.

Hora had designed his watches such that he could put together subassemblies of about ten components each. Ten of these subassemblies could be put together to make a larger subassembly. Finally, ten of the larger subassemblies constituted the whole watch. Each subassembly could be put down without falling apart.

The first observation was that complex systems will evolve from simple systems much more rapidly if there are stable intermediate forms present. This observation was influenced by Simon's parable of the two watchmakers (Simon, 1990). Simon's parable (see frame) demonstrates how, in dynamic and demanding environments, the chances of emerging and surviving for systems composed of suitable subsystems are vastly superior to systems composed from basic elements without stable intermediate states or subsystems.

Indeed, bounded or limited rationality – finite brainpower, bounded information processing, and communication capacity – puts a ceiling on the speed at which elements can be integrated to build a system. When these elements are small, more time and effort is required to build a system of a given size. In combination with a dynamic environment, the resulting system is completed too late to be effective or competitive. Systems built from larger building blocks are superior in environments that emphasize adaptation speed over the theoretical possibility of superior ultimate performance.

In Koestler's words, larger holons (holonic systems) are constructed from stable intermediate forms, also called holons, where this composition repeats in a self-similar manner until the constituents become simple. Such holonic systems are more likely to emerge and survive in dynamic environments than systems that would ultimately be superior but take too much design effort and development time. The latter systems simply are too expensive and, most importantly, obsolete long before they become operational.

The second observation was that although it is easy to identify subwholes or parts, "wholes" and "parts" in an absolute sense do not exist anywhere. The term *holon* was proposed to describe the hybrid nature of subwholes/parts in real-life systems. Holons simultaneously are self-contained wholes to their subordinated parts, and dependent parts when seen from the inverse direction. Or put more simply, a holon is something that is whole in itself as well as part of a greater whole. Koestler called this behavior the "Janus effect."

Koestler also points out that holons are autonomous self-reliant units, which have a degree of independence and handle contingencies without asking higher authorities for instructions. At the same time, these holons are subject to control from higher authorities. The first property emphasizes that holons are stable forms and can cope with disturbances. The second property highlights that holons are intermediate forms, providing the proper functionality for the larger whole.

According to Koestler, a holonic system or holarchy is then a hierarchy of self-regulating holons that function (i) as autonomous wholes in supra-ordination to their parts; (ii) as dependent parts in subordination to control at higher levels; and (iii) in coordination with their local environment.

Simon's main goal is to explain why our universe is dominated by systems exhibiting hierarchical structure in a loose sense – which Koestler calls "holarchies" to distinguish them from strict hierarchies (typical for rigid man-made artifacts). Koestler makes a first attempt at characterizing these suitable subsystems, which have to survive the dynamics of the environment better and longer than the overall system, as illustrated in the watchmakers' parable. Koestler calls this ability to survive the autonomy of the subsystem or holon, whereas the contribution of the subsystem to the overall system is called the cooperativeness of the holon. The autonomy gives the holon the capacity to cope with changes, uncertainty, and disturbances in its environment.

The first law of the artificial states that flexible hierarchies will dominate when and where the first three axioms hold; holonic systems will be the superior choice. Here, "flexible" means that these hierarchies adapt their composition over time by replacement of subsystems at appropriate hierarchical levels. When these three axioms fully apply, holonic systems will be your only possibility because anything else is eliminated by ruthless and fierce holonic competitors.

However, Koestler's analysis concerning the stability of intermediate forms (e.g., their autonomy) fails to qualify as a law. The stability and availability of suitable intermediate forms is dominated by the second law (i.e., membership of strong autocatalytic sets), which is discussed below. Note that autonomous contingency handling is likely to induce and reinforce such membership but it is not the sole nor a dominating factor in this respect.

LAW 2: AUTOCATALYTIC SETS – CRITICAL USER MASS

An *autocatalytic set* is a collection of entities, each of which can be created "catalytically" by other entities within the set, such that as a whole, the set is able to catalyze its own production and expansion. In this way, the set "as a whole" is said to be "autocatalytic."

The second law of the artificial states that members of strong autocatalytic sets have decisive competitive advantages over nonmembers. In competitive environments, resources will go (almost exclusively) to members, rendering nonmembers to become extinct through starvation.

Autocatalytic sets were originally and most concretely defined in terms of "molecular entities" but have more recently been metaphorically extended to the study of systems in sociology and economics. Here, we extend them to (complex) artifacts, designed and produced by humans. Autocatalysis reveals to be more fundamental than autonomy for the emergence and survival of systems, which eventually become subsystems in a larger system. Autonomy and adaptability are secondary properties of autocatalytic sets and their members, useful to maintain and increase autocatalysis in a dynamic and changing environment.

Originally, the concept of an autocatalytic set (Hordijk, 2013) served to enhance the probabilities in the standard biologist's theory that life emerged by chance when energy pulses stroke a pool filled with organic molecules. Unfortunately for this standard theory, the smallest life forms still are so big that combining their basic molecules by chance is as likely to happen as "a group of monkeys typing Shakespeare's oeuvre by pure coincidence."

These calculations change drastically, however, if combinations of molecules are formed into sets that are catalysts for themselves. If energy pulses arrive at a sufficiently high frequency, the autocatalysis implies that the pool rapidly becomes filled with members of such autocatalytic sets (exponential growth until raw material becomes scarce). The omnipresence of such set members also means that they become the building blocks for larger molecular combinations, among which the autocatalytic set members will dominate again.

In view of the first law of the artificial, membership of sufficiently strong autocatalytic sets delivers an exponential competitive advantage. Our holons will be members but not all members will be suitable subsystems. Note also that being a suitable subsystem provides (indirect) membership to the autocatalytic sets of the larger system. Indeed, the first and second law of the artificial mutually reinforce each other.

In the biologist's theory, this interaction creating larger entities (first law) among which the autocatalytic set members are favored (second law) can be repeated until life forms emerge. If this theory is correct, the dominating life forms should be members of autocatalytic sets themselves. And indeed, mice, rabbits, weeds and insects all provide strong empirical evidence supporting the theory (i.e., their reproduction, sexually and otherwise is autocatalysis indeed). Autonomy, adaptability, manipulating the environment, etc. are secondary properties of the more complex life forms (including humans) that mainly increase the intensity of the autocatalysis in favor of the own set.

To translate the above to the domain of holonic systems and mechatronic systems/societies, it is necessary to identify the relevant autocatalytic sets for manmade artifacts and software systems in particular. These sets are not situated in artificial worlds inside computer platforms serving to investigate artificial life. The relevant autocatalytic sets comprise both software and humans (i.e., software users and developers).

Successful software systems belong to two kinds of autocatalytic sets: (i) the *economic autocatalytic set* – successful software represents economic value to its users and thus mobilizes the economic means for software developers to maintain, adapt, and enhance this software; (ii) the *information feedback autocatalytic set* – successful software attracts users providing feedback on its shortcomings and merits. This information, in combination with the economic means, allows the developers to maintain, adapt, and enhance the software such that it remains competitive.

The above implies that (software) system designers have to account for more than just the technical dimension to be successful. Sufficient users (and their payments) and sufficient diversity in the user community (and its feedback), relative to the competitive pressures, are necessary for emergence and survival. Most importantly, such successful members of autocatalytic sets are the (only) systems that may become the subsystems in larger, more sophisticated systems.

The second law of the artificial translated toward (software) artifacts states that the – potential for – critical user mass is the principal factor in achieving membership of strong autocatalytic sets. This user mass must be high, relative to the complexity of the artifact, in both the economic and information feedback aspects.

Ceteris paribus, software systems with the intrinsic ability to serve more users or a more diverse community of users will have an edge over the competition since their autocatalysis is stronger. Note that software quality and functionality levels are likely to improve significantly through the above types of autocatalysis. Personal experience with systems experiencing low levels of autocatalysis has provided the authors with strong empirical evidence supporting this statement.

LOCK-IN

The previous section depicts how positive feedback is instrumental in structuring worlds such that larger and more complex systems emerge and survive. This section introduces a negative aspect of such positive feedback: *lock-in* into early solutions. Indeed, the competitor achieving autocatalytic growth first generally ends up dominating its world. Competing and faster autocatalytic processes only have a short time window to overtake a competitor that started earlier. Soon after his start, the first one to achieve autocatalysis will have consumed or acquired most resources – needed to grow and sustain – which are also needed by the other competitors.

In other words, systems, in which autocatalysis occurs, may evolve along multiple trajectories where the selection among these trajectories strongly depends on which autocatalytic process kicks in the earliest. Because it is an exponential process, the autocatalytic set rapidly exhausts the available "raw material," effectively eliminating the opportunity for other autocatalytic processes sharing "material requirements" to start at all.

A well-known example is the "VHS versus Betamax and Video 2000 war" in television tape recorders. VHS was the first to attract a critical user mass and, because it was the only competitor that was certain to provide access to a sizeable market for content providers, it deprived the competing technologies from content, in spite of their technological superiority. Here, the exponential nature of autocatalysis prevents better solutions from emerging; a good enough alternative that arrives earlier wins and locks the world into staying with its first choice.

Relevant for holonic systems development is the poor level of suitability and adaptability of the available systems (members of autocatalytic sets) from which larger systems have to be developed. Today's systems were developed with specific user requirements in mind, and the world locks into those early solutions. Those early solutions incorporate many design choices that prevent the creation and maintenance of other and larger systems at some later time. Likewise, larger systems can be created but lock-in into design choices from the early solutions results in poor performance. This often is referred to as "legacy" in its negative connotation.

Overall, lock-in is desirable when it simplifies the world by reducing the options and alternatives that have to be taken into account. Lock-in is undesirable when a dominating system incorporates highly unfortunate design choices, which unfortunately are commonplace because being first (and just good enough in the short run) is more important than being well designed.

Relevant for design for the unexpected (D4U), lock-in denies our society the ability to make progress incrementally through a sequence of short-term good-enough solutions, which are improved and combined as time progresses. This approach rapidly runs into the ground when early design choices, which have become hard-to-undo, prevent the desired improvement. D4U specifically aims at developing solutions that do not make problematic design choices or render it easy to undo problematic design choices. In other words, it aims to create a legacy-free base concerning lock-in.

LAW 3: STEERING WITHOUT CENTRALIZATION

When a competitive environment favors larger systems (i.e., superior size brings a competitive advantage), systems will grow until centralized coordination and control is no longer feasible (because of bounded rationality). The following anecdote (Kurtz and Snowden, 2003) illustrates in a small-scale setting what happens:

> A group of West Point graduates were asked to manage the playtime of a kindergarten as a final year assignment. The cruel thing is that they were given time to prepare. They planned; they rationally identified objectives; they determined backup and response plans. They then tried to "order" children's play based on rational design principles, and, in consequence, achieved chaos.
>
> They then observed what teachers do. Experienced teachers allow a degree of freedom at the start of the session, then intervene to stabilize desirable patterns and destabilize undesirable ones; and, when they are very clever, they seed the space so that the patterns they want are more likely to emerge.

At larger scales in knowledge-intensive systems, a similar situation transpires. Regarding the upward information flow, the small children lacked the required intellectual skills. In the scaled-up situation, information overload occurs when the information is gathered and fed into higher levels. Moreover, bounded rationality prevents adequate condensing of this flood of information. Concerning the downward information flow, the condensed nature of information flowing down – as it is generated by higher levels with a reduced information-processing capacity – implies a need for expansion and interpretation. This will distort the information.

Worse, the down-flowing information must be disregarded whenever the loss of information in the upward flow causes this down-flowing information to be irrelevant and/or wrong. When it is not disregarded, centralized coordination and control become a liability and harm the overall system. This is also observed at on a smaller scale when the upper level lacks basic knowledge in a specialist's domain.[1]

[1] Everyone knows that cars and trucks should travel on the same side of the road. But in lesser-known domains, policy makers are likely to decide that tomorrow – as an analogy – cars shall drive on the other side of the road and, when all goes fine, trucks will change to this new side three weeks later. Actually, knowledgeable opponents of this change may propose and encourage such a flawed coordination and control scenario.

In other words, classical command and control hierarchies become ineffective and even counterproductive at some point when systems grow larger. A leadership that feels responsible for the overall system performance and considers it their duty to steer it in a centralized top–down manner becomes a liability. These large systems need leaders who aim to facilitate and who primarily aim not to harm performance on the lower levels – that is, they assume the lower levels aim to perform without the need for explicit command whereas the higher levels cannot match the combined information processing power of the lower levels.

The third law of the artificial states that when competition favors larger systems, the top levels of competitive systems cannot be effectively controlled in a centralized manner. Noncentralized designs to steer these large systems exist but much still needs to be invented and remains to be discovered. Here, the social sciences and humanities may find some challenges to address.

Other coordination, influencing, and guiding mechanisms are needed, exists and can be designed. Indeed, in a competitive environment, systems are likely to benefit from coordination and guidance, which needs to emerge rather than resulting from higher levels imposing it. The following fundamental research experiment in robotics illustrates that it is possible to design such emergent coordination; on purpose, it severely limits communication producing a situation that normally only occurs in much larger systems:

In a room, a number of mobile robots are moving around. These robots have no means for communication and are only capable of three actions: moving around the room, recharging their battery in a docking station, and pushing a button on the other side of the room. The battery charging station is only active/powered when this button is pushed. A robot cannot push the button while docked in the charging station. The objective of these robots, employed to study machine-learning mechanisms, is to spend as much time roaming around the room as possible.

A successful strategy consisted of robots pushing the button with a frequency and duration that is close to what they perceive to be the overall average while deviating slightly in the direction that the individual robot deems desirable. As a group behavior, this would steer the overall system toward good performance while it prevented the adaptation mechanisms in the robots from exploiting overly generous robots, which in turn avoided the need for detection and punishing mechanism for profiteering/abusing robots.

The above illustrates coordination without a centralized control. Although academic, it already reveals how to handle socioeconomic situations in which a small player may wish to influence the overall system without risking to be exploited (i.e., for being too nice and gullible) or exclusion (i.e., for being more royalist than the king).

Our body of knowledge on the design of such mechanisms in real-world complex cases remains embryonic today. In view of the design theory in this book, ongoing research focuses too much on the decision mechanism design (e.g., how much more shall a robot push the button) and not enough on an environment to facilitate and trigger desirable behavioral patterns in the system (cf. the experienced and clever teachers in the above kindergarten).

A well-known example in nature to achieve coordinated behavior without a centralized control is food foraging in (many types of) ant colonies. These ants use chemical trails (pheromones) to guide other members from their colony to food sources. The brilliant element in this design is not the tuning of the performance-determining parameters but the manner in which the ants reuse the environment itself to cope with (the complexity and dynamics of) this environment.

In other words, when designing noncentralized coordination and control, the main achievements will reside in structuring the environment whereas performance-determining decision-making mechanisms and their tuning methods are more or less instances of a limited number of generic designs. Rather than providing elements that take decisions, the proper design must deliver an environment that

- enhances visibility (situational awareness), allowing the decentralized players to take informed decisions and actions.
- allows for larger deviations from the average/group behavior without repercussions. For instance, the designer may create an environment in which good deeds are rewarded, or at least remain unpunished.
- provides mechanisms to assess the effect of alternative courses of action under consideration. The actors shall have a kind of radar or crow's nest to take better-informed decisions.
- provides mechanisms to make the above a collective capability, avoiding that individuals have to second-guess what the impact of others will be. In such an environment, individual actors shall never conclude that "if I had known this earlier, I would have acted differently," particularly in situations where the desirability and performance of their actions depends on what other actors do (e.g., whether to stay longer in the office to avoid rush hour).

- provides mechanisms to make commitments among the participants involved. This allows a participant to self-command (individually or collectively) and, simultaneously, get the benefit of planned action when this participant is sharing what she or he is intending to do (including with a specified commitment). This renders the collective more predictable without needing a central coordination taking decisions.
- seeds the environment with information to guide decisions without imposing them. This will enable don't-care decisions to be compatible, initiate search for solutions from promising starting points, and may make the system behavior simpler and more predictable (i.e., converge faster and converge in a more predictable manner). The discussion on staff holons in Chapter 5, enlightens this point.

Summarizing, coordination and control in large complex-adaptive systems will not be centralized at the higher levels, simply because of limited rationality. Nevertheless, significant system design effort remains possible and beneficial. This design effort shall primarily focus on seeding the environment to trigger desirable collective and individual behavior, including simplicity, predictability, and effectiveness. The decision-making elements in this setting are mostly instances from generic designs that need nonstop tuning within this environment. The performance of these mechanisms is bounded by the environment's offerings. A top-performing ship's captain without radar is no match for a capable captain with effective radar. Moreover, the design of this environment needs to account itself for the theoretical insights discussed elsewhere in this book (holonic, autocatalytic, etc.).

LAW 4: COLLECTIVE IMAGINATION AND PROACTIVENESS

The most significant word in complex-adaptive systems is "adaptive"; this is what many research communities have barely covered or not at all. Vaccination is a well-known mechanism that exploits this phenomenon on the individual level. On the other extreme, whole societies adapt. In warfare, an army attempting to repeat a successful strategy on the same enemy, which lost the battles in the past but survived, is unlikely to repeat the corresponding success. The battles in the past are likely to have "vaccinated" this enemy. But even in noncompetitive situations, past behavior and statistical data may be insufficient to enable an effective proactiveness.

Consider an intelligent transport system that monitors traffic density and that combines historical data with models to predict congestion. When this information is shared with a limited number of paying customers, these privileged participants will benefit and avoid being stuck in traffic jams. When the same information is made available to everyone, the resulting changes in behavior of the participants will invalidate the historical data and the models. Indeed, the participants react to the prediction in sufficiently large numbers and extent to cause the prediction to be wrong (and they get stuck on the routes avoiding the predicted congestions).

The above illustrates how decisive the adaptive nature of elements in especially large systems can be. As discussed above, these systems are too large and complex for centralized planning, scheduling, or control. At the same time, fully decentralized designs are unable to "imagine" what will happen (where this ability is vital for proactiveness). For instance, designs in which "agents" have a closed-formula objective or utility function (exclusively based on information owned by the agent) will be myopic as this function fails to grasp this decisive impact of future interactions. Nonetheless, this ability to imagine, predict, and simulate what will happen when selecting a course of action is vital for system performance and service levels. Proactiveness needs this ability to assess future performance to be effective.

In the present discussion, it is inherently impossible to derive such predictions based on past behavior at the level of an overall system. However, sufficiently accurate models of smaller components do exist (it suffices to descend until the system components become sufficiently small, simple and stable). In the applications addressed in this book, it is equally possible to model the courses of action under consideration. Moreover, the models are executable software allowing to virtually execute such a course of action.

The above capabilities are used to collectively imagine – in a decentralized design/manner – what will happen. To this end, the intended courses of actions of all the participants are virtually executed on models of the resources (e.g., cars driving given routes). Because the intentions are virtually executed to generate a collective prediction, the accuracy of the prediction will depend - among others - on the respective commitments to these intentions.

Regular refreshing, by repeating this virtual execution of the intentions, allows this collective prediction to stay up-to-date. In addition, the virtual execution of a course of action under consideration (not an intention) serves to explore and collect information used to decide on introducing a new course of action or changing an existing course of action. Further discussion can be found in the section on DMAS in Chapter 5.

Summarizing and generalizing, proactiveness and its benefits require the ability to imagine what will happen when selecting a course of action, including the impact of future interactions:

The fourth law of the artificial states that effective proactiveness – and its benefits – requires the ability to imagine what will happen when selecting a course of action. This imagination must include the significant impact of future interactions. Accounting for such future interactions requires a collective imagination. This includes the impact of the accessibility of this imagination, collective or otherwise.

Note that the construction of such collective imagination offers an opportunity to enjoy the best possible prediction of the future: "the best manner to predict the future is to create the future." Indeed, while participating in the creation and maintenance of such a collective imagination, the participants are – to a given extent – jointly creating their future.

CONCLUSION AND REMARKS

Four observations of key properties of our world, called axioms, give rise to four statements, called laws of the artificial, about the man-made world. In order to be successful, the designer of complex artifacts should be aware of the existence of these laws in his or her design efforts, in the same way as he/she should recognize the laws of nature.

In Chapters 5–7, these laws of the artificial will serve as the major guideline for designing decentralized holonic control architectures for mechatronic societies.

ABBREVIATION

DMAS Delegate multiagent system

REFERENCES

Hordijk, W., 2013. Autocatalytic sets: from the origin of life to the economy. BioScience 63, 877–881.
Koestler, A., 1967. The Ghost in the Machine. The Macmillan Company, Hutchinson, UK.
Kurtz, C.F., Snowden, D.J., 2003. The new dynamics of strategy: sense-making in a complex and complicated world. IBM Syst. J. 42 (3), 462–483.
Simon, H.A., 1990. The Sciences of the Artificial. MIT Press, Cambridge, MA.

CHAPTER FIVE

Holonic Manufacturing Systems

Paul Valckenaers*, Hendrik Van Brussel**
*Faculty of Engineering Technology, KU Leuven
**Faculty of Engineering Science, KU Leuven

This chapter presents the earlier results in our development of a holonic manufacturing systems framework, with a clear emphasis on manufacturing and logistics. The insights discussed in the previous chapters were used as precious guidelines and as a check for soundness when comparing options. In particular, design for the unexpected (low and late commitment) and maximizing the potential for achieving critical user mass were important factors influencing the choices made during the development activities.

These earlier developments concentrate on manufacturing execution systems (MESs). This involves some serious challenges – unpredictable production processes (duration, outcome), heterogeneity of products and equipment, etc. – in combination with a competitive environment. When decisive for the competitiveness, an MES must cope with the most exotic properties and behaviors of a production system. Indeed, design for the unexpected becomes a necessity.

Conversely, these developments assumed that a single organization decided what software would be installed and used. In this respect, a closed-world assumption was made, limiting the developed concepts to "closed systems." Chapter 6 presents more recent, consolidated successors of the developments discussed here. These successors relax and abandon this closed-world assumption and have become agnostic concerning the application domain enabling the formation of a humane mechatronic society.

HOLONIC SYSTEMS

As shown in Chapter 4, the Law of Holonic Systems, derived by Koestler and Simon, is based on two observations:

- Complex systems will evolve from simple systems much more rapidly if there are stable intermediate forms present.
- Although it is easy to identify subwholes or parts, "wholes" and "parts" in an absolute sense do not exist anywhere.

Design for the Unexpected. http://dx.doi.org/10.1016/B978-0-12-803662-4.00005-9

The "holon" concept was proposed to describe the hybrid nature of subwholes/parts in real-life systems. Holons simultaneously are self-contained wholes to their subordinated parts, and dependent parts when seen from the inverse direction.

Koestler points out that holons are autonomous self-reliant units, which have a degree of independence and handle contingencies without asking higher authorities for instructions. At the same time, these holons are subject to control from higher authorities. The first property emphasizes that holons are stable forms and can cope with disturbances. The second property highlights that holons are intermediate forms, providing the proper functionality for the larger whole.

According to Koestler, a holonic system or holarchy is a hierarchy of self-regulating holons that function

- as autonomous wholes in supraordination to their parts;
- as dependent parts in subordination to control at higher levels;
- in coordination with their local environment.

Shortly after its publication, Koestler's book (Koestler, 1967) went into oblivion. It was rediscovered by Japanese manufacturing scientists, around 1989, in their quest, assigned by MITI (Ministry of International Trade and Industry), to find the appropriate structure of the factory of the 21st century. They declared the holonic system concept as the basis for the factory of the future. It was further developed and translated for manufacturing in the framework of the worldwide IMS (Intelligent manufacturing system) project, launched by MITI under the impulse of Professor H. Yoshikawa of the University of Tokyo. The authors of this book participated, from 1993 onward, in a worldwide consortium, called Holonic manufacturing systems (HMS). It was there that the seeds of this book were sown.

Based on the concepts of Koestler, a new form of manufacturing systems, called *holonic manufacturing systems*, emerged. This new paradigm had the ambition to provide an answer to the shortcomings of earlier factory control systems that led to the failure, and ultimate demise, of the then-prevailing computer integrated manufacturing (CIM) paradigm.

Previously, the control architectures of these CIM systems had evolved from centralized via hierarchical to heterarchical architectures (Dilts et al., 1991). A *control architecture* determines the interrelationships between the various system components and allocates the different decision-making responsibilities (e.g., part routing and resource allocation) to specific control components.

Centralized Control

Centralized control architectures are characterized by a central computer that performs all planning and information processing and registers the activities of the whole manufacturing system. This way, overall system status information can be easily retrieved from a single source. This ability to access complete global information also makes optimization a more realistic expectation. However, centralized architectures tend to have a poor responsiveness, reliability, modifiability, and extensibility.

Hierarchical Control

The shortcomings of centralized control resulted in the development of *hierarchical control architectures*. These architectures introduce "levels" of control that have a specific functionality and are organized in a top–down structure. There are strict master–slave relationships between the levels. Control decisions are operated top–down, whereas status reporting operates bottom–up. The benefits of these architectures include optimal behavior when everything keeps going right, fast response times, gradual implementation, redundancy, and limited complexity of individual control modules. Despite these advantages, there are also many drawbacks. The rigid structure makes it very difficult to make unforeseen modifications and the increased coupling between the modules adversely affects modifiability, extensibility, and fault-tolerance. As low-level modules have to consult higher levels in the hierarchy in case of a disturbance, the system's reactivity to disturbances is weak. Moreover, global decision-making is often based on obsolete information.

Heterarchical Control

Heterarchical control architectures are characterized by a flat structure. They consist of distributed locally autonomous entities that cooperate with each other directly (without the master–slave relationship from hierarchical architectures to make global decisions). This local autonomy requires that global information is minimal or even nonexisting. Advantages are enhanced modularity, reduced coupling between the modules, and increased robustness against disturbances. A major disadvantage is the low predictability of heterarchical architectures, as it is difficult to operate according to a predefined plan. Moreover, there is no global optimization possible and, consequently, a high performance cannot be guaranteed.

Holonic Manufacturing Systems

Holonic manufacturing systems (Babiceanu and Chen, 2006) were put forward as it was realized that neither centralized, hierarchical nor heterarchical control systems could face the challenges the manufacturing world was confronted with. Holonic control systems try to combine the high and predictable performance promised by hierarchical systems with the robustness against disturbances and the agility of heterarchical systems. To avoid the rigid structure of hierarchical systems, holonic manufacturing systems provide autonomy to the individual holons. This allows the control system to respond quickly to disturbances and to reconfigure itself to face new requirements. In order not to ban all hierarchy, which is essential to master complexity, holons work together in "loose" hierarchies. Such a hierarchy is different from a traditional hierarchy in that

- holons can belong to various hierarchies,
- holons can form temporary hierarchies, and
- holons do not rely on the correct functioning of the other holons to perform their tasks.

To develop a holonic manufacturing system, the concepts developed by Koestler were translated into a set of appropriate concepts for manufacturing (Babiceanu and Chen, 2006). The HMS consortium developed the following list of definitions to help understand and guide the translation of holonic concepts into a manufacturing setting:

Holon

A holon is an autonomous and cooperative building block of a manufacturing system for transforming, transporting, storing, and/or validating information and physical objects. The holon consists of an information-processing part and often a physical processing part. A holon can be part of another holon.

Autonomy

It refers to the capability of a holon to create and control the execution of its own plans and/or strategies.

Cooperation

It is the process whereby a set of holons develops mutually acceptable plans and executes these plans.

Holarchy

A system of holons that can cooperate to achieve a goal or objective is a holarchy. The holarchy defines the basic rules for cooperation of the holons and thereby limits their autonomy.

As a holon is an autonomous entity that cooperates with other holons to achieve its goals, the *multiagent systems* paradigm seems very appropriate to implement holonic manufacturing systems. There are however two important differences between "holons" and "agents." First, a holon can contain one or more other holons, whereas an agent is not composed of other agents. Second, while agents are pure software entities, holons can include both hardware and software parts. Conceptually, multiagent systems[1] are a most natural choice to implement holonic systems. Surprisingly, Erlang/OTP – an open source telecom technology, which is downloadable from www.erlang.org – proved to be the most appropriate software development and deployment technology for the research addressed in this book. The programming tools and languages from the multiagent community were lacking in maturity and critical user mass; they were underpowered where it hurts, whereas their strengths barely qualified as nice-to-have.

THE PROSA REFERENCE ARCHITECTURE (Van Brussel et al., 1998)

This section addresses the structure of a holonic manufacturing system or, in other words, its architecture. It addresses the structure of the system of holons (the holarchy), not the internal structure of individual holons.

A reference architecture describes the mapping from various functionalities, which cooperatively solve the problem, onto software components and the data flows between these components. A reference architecture is not an architecture in itself but can be used as the basis for designing the system architecture for a particular system. For instance, the ADAptive holonic COntrol aRchitecture (ADACOR), can be considered as an instantiation of the product-resource-order-staff architecture (PROSA) reference architecture (see Chapter 8: Work by Others).

Reference architectures are used in a specific (mature) domain and arise from experience (Wyns., 1999). The reference architecture "Gothic cathedrals" is the collection of knowledge and skills, acquired by the medieval guilds, to build Gothic cathedrals. The cathedral of Chartres, with its exquisite architecture, is an impressive instantiation of that reference architecture.

PROSA was originally developed for the manufacturing domain and, based on experience in this domain, special attention was paid to (Wyns., 1999) (i) separating the essential elements, which are generic, from

[1]See Appendix II for a discussion of some basic concepts of multi-agent systems.

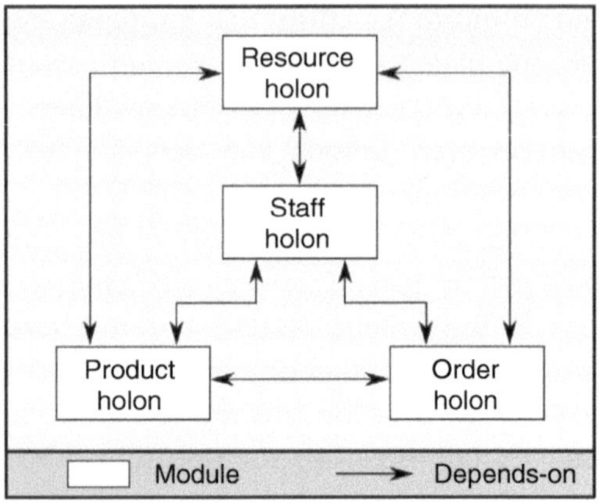

Figure 5.1 ***Overview of the PROSA Reference Architecture.*** **Each rectangle represents one of the holon types in PROSA; the arrows between the holons represent their interactions.**

the optional elements, which can be domain specific (the latter are called "plugins"); (ii) separating the structural aspects from the functional (algorithmic) aspects for resource allocation and process planning; (iii) separating resource allocation aspects and process specific aspects; and (iv) enabling the incorporation of legacy systems, or the introduction of new technology.

The PROSA reference architecture is developed in accordance with the holonic-manufacturing paradigm. The basic components are holons, and the architecture describes the responsibilities of the various holons and their interactions. The acronym PROSA stands for *product-resource-order-staff architecture* and refers to the different types of holons. Three basic types of holons can be distinguished: product holons, resource holons, and order holons. Staff holons complete the set of PROSA holons.

Each basic holon represents a separate concern in the application domain: process planning, resource allocation, and logistics management, respectively. The basic holons can be aggregated into larger holons and specialization can be used to structure them. Staff holons are optional and can be added to provide the other holons with expert knowledge or to incorporate legacy systems. Figure 5.1 shows a module decomposition view[2] of

[2]A module is an implementation unit of software that provides a coherent set of responsibilities. A module decomposition view describes the organization of the software as modules and submodules and shows how responsibilities are divided across these modules.

the holonic reference architecture (Verstraete et al., 2008). The depends-on relationships between the holons indicate that the various holons share data with each other.

The source of inspiration for PROSA was the kernel of a modern computer operating system. It recognizes resources (the microprocessor), know-how (the code segment), and execution (the data segment), which map, respectively, on resource, product, and order holons. The staff holon has no counterpart in computer operating systems. The following paragraphs elaborate on the four different holon types.

Resource Holon

A *resource holon* corresponds to a resource in the underlying domain (equipment, infrastructure elements, personnel). In a logistic context, for instance, this means that all transport means (trucks, freight trains, cargo aircraft, etc.) and material handling equipment (forklift trucks, conveyors, automated guided vehicles, etc.) will be represented by a resource holon. There will also be resource holons for other entities that are scarce and have to be shared (e.g., dock doors, pallet racks, and floor space). Note that a logistic execution system (LES), like MES, can be developed under a closed-world assumption.

Each resource holon comprises the physical resource, together with a software part that controls this resource. It offers knowledge about processing capacity and processing functionality to the other holons and organizes and controls the usage of the physical resource. More concretely, a resource holon has the following responsibilities:

- *Reflection of reality:* A resource holon reflects its corresponding physical resource; that is, it contains information about the current state of the resource and expected future states. It should keep the reflection of the resource state synchronized with the actual resource state. Moreover, the holon has knowledge about the dynamic behavior of the physical resource and can answer what-if questions (e.g., what will be the arrival time of a truck if it departs at a certain time).
- *Information provision:* A resource holon should be able to provide resource-related information to the other holons. This includes process information (e.g., possible operations), information about the local topology (which other resource holons this holon is logically connected with), and about possible constraints (e.g., truck capacity, maximum cargo weight, etc.).
- *Maintaining a local schedule:* Each resource holon owns an agenda in which its future tasks/operations are recorded, based on requests from

order holons. This can be seen as a reservation service that keeps track of the availability of the resource over time. Each operation to be processed by the physical resource needs to be reserved on beforehand in this local schedule. To cooperate with the delegate multiagent system (DMAS) pattern (see further), the local schedule is implemented as a (virtual) blackboard structure and applies an "evaporation-refresh" mechanism.

- *Managing its local schedule:* The resource holons have local authority on how they organize (sequence or schedule) the various operations (from order holon requests), for instance, by applying priority or batching rules. This local decision-making is resource-specific and mainly depends on the performance settings of the resource.
- *Virtual execution:* This responsibility is a service for the order holons who can request information on the virtual outcome of an operation (e.g., quality and end time). Based on the local schedule (to decide how the operation can be fitted in between the already reserved operations) and its what-if functionality (to virtually execute the operation), the resource holon is able to provide accurate information to the order holon.
- *Controlling the resource:* A resource holon controls the real-world resource by starting and stopping the (scheduled) operations and by monitoring the execution.

Several resource holons can be clustered together to form a bigger resource holon with its own identity. An example of such an *aggregated resource holon* is shown in Figure 5.2. A cross-dock holon consists of a temporary storage holon, one or more forklift holons, and one or more dock door holons. The granularity of the aggregation will depend on the application and on the need of explicitly allocating these resources. For instance, it can be required to explicitly consider the forklift driver, and to see the forklift holon as an aggregate of a forklift truck holon and a forklift driver holon.

Specialization can be used to differentiate between the different kinds of resource holons. Figure 5.3 shows an example of such a specialization.

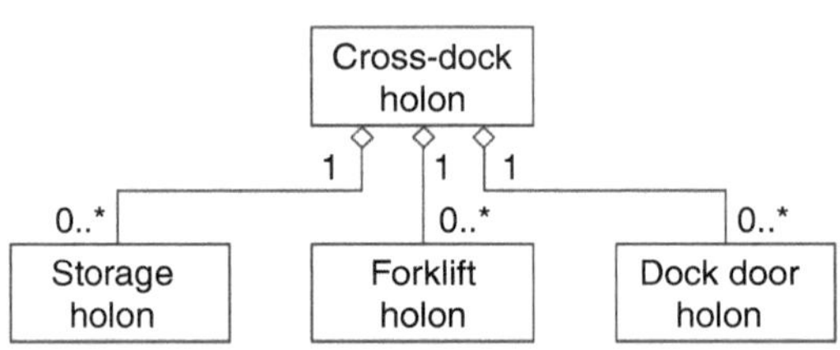

Figure 5.2 ***Example of an Aggregated Resource Holon.*** Each rectangle represents a resource holon; the connections indicate aggregations.

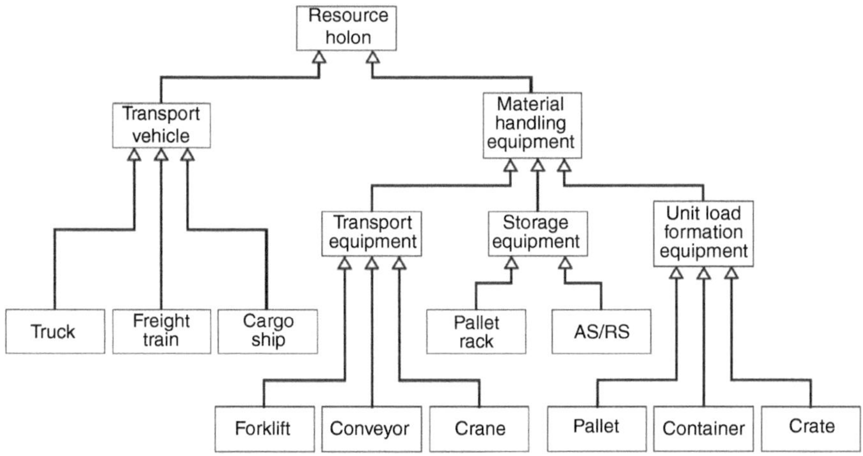

Figure 5.3 ***Example of Specialization of Resource Holons.***

Transportation vehicle and material handling equipment are both resource holons. Transport equipment, storage equipment, and unit load formation equipment are all kinds of material handling equipment. Pallet racks and automated storage and retrieval systems (AS/RS) are examples of storage equipment.

Product Holon

A *product holon* corresponds to a task type or order type. To accomplish the task or to fulfill the order, a process (a sequence of operations) has to be executed. The product holon contains the knowledge on how instances of a specific task type (represented by order holons) can be executed by the resources, that is, which operations are required to accomplish the task correctly and qualitatively. For instance, to deliver a package, the product holon knows that this package has to be picked up, transferred to the cross-dock, consolidated, and eventually brought to its destination. The product holon also has information about constraints on or process parameters of these operations. For instance, if the package contains refrigerated products, the holon knows the allowable temperature range to which the package can be exposed during transportation. Note that a product holon only holds information about its order type, and not about individual order instances. The main responsibilities of a product holon are as follows:

- *Maintaining process knowledge:* The product holons hold the necessary process knowledge to realize instances of their type. This includes, amongst others, process plans, process parameters and quality requirements. The product holons are responsible for keeping this information consistent and up-to-date, for instance, if new operations are offered by (new) resources.

- *Determination of operation options:* A product holon informs the order holons about all possibilities for their next operation. Indeed, after completion of an operation, the order holon needs information about its next operation in order to accomplish its task. In its simplest form, the process plan is linear and the product holon supplies the order holon with only one possibility. In general, multiple options are possible and these alternatives depend on the current state of the order holon, that is, based on the outcome of the previous operation. For instance, if the quality of an operation is insufficient, the product holon can decide that the operation has to be redone. As another example, if a package with refrigerated products is exposed to too high temperatures during its transportation, the holon can decide that the package should be disposed of instead of being delivered. The selection of one operation out of the available options is the responsibility of the order holon itself.
- *Process information provision:* Just before a selected operation should start on a resource, the resource holon needs to know the desired process parameters (e.g., temperature during transportation in a refrigerator truck). The product holon is responsible for providing this process information to the resource holons. By providing this information just before the operation starts, the product holon is able to take the latest state of the order holon into account.

Similar to resource holons, product holons can be combined into an aggregated product holon that represents the combination of the corresponding process plans. For instance, the aforementioned product holon responsible for delivering a package could consist of three product holons: one to transport the package to the cross-dock, one to process the package inside the cross-dock, and one to deliver the package to its destination. The aggregated holon delegates some of its responsibilities to these subholons, but is still in charge. This aggregation limits the complexity of the holons and allows an easy introduction of new product holons by combining other product holons.

Order Holon

An *order holon* corresponds to a task (instance) or order (instance) that needs to be executed, for example, the delivery of a package. Each order holon is closely linked to the product holon representing the corresponding task or order type. Although a product holon can be linked to multiple order holons, each order holon will be associated with only one product holon. The order holon is responsible for handling the required resource allocations in order to accomplish the correct execution of its task. To this end, the order holon consults its corresponding product holon to find out which operations it needs to perform and searches for the proper resources and time slots to execute these operations.

In a logistic context, the order holons can often be associated with physical entities, that is, the freight units that have to be transported (e.g., pallets). The order holon then consists of this real-world entity, together with a software part that controls the execution of the corresponding task. In a manufacturing context, an order holon might correspond to a number of products that have to be produced by a certain due date.

More concretely, an order holon has the following responsibilities:

- *Reflection of reality:* An order holon reflects the order instance, that is, it contains information about the current state of the order and the corresponding physical entity. This includes, for instance, the location of the order, the current operation being processed, the resource performing this operation, etc. The order holon is responsible for keeping the reflection of its state up-to-date with the actual state.
- *Searching solutions:* The order holons search for solutions to execute their tasks. During their search, the order holons will consult their product holons to know the required operations and will virtually execute these operations (by using the virtual execution service of the resource holons) to check for resource availability.
- *Intention selection:* Each order holon evaluates the solutions it has found and chooses the most attractive solution (according to its performance measure) to become its intention.
- *Reserving its intention:* The order holon then informs the other holons about its intention by making the necessary reservations (future allocations) at the involved resource holons. As these reservations evaporate after a certain time, the holon has to confirm its reservation at regular time intervals.

Furthermore, order holons can be aggregated into an aggregated holon. For instance, several orders corresponding to freight that has to be transported can be aggregated into one batch in order to be transported together by a truck. Over time, an order can be part of multiple batches for different transport operations.

Staff Holon

The three basic types of holons can be assisted by one or more staff holons. These holons can provide the other holons with expert knowledge about certain aspects of their decision-making. For instance, a staff holon can give information on which containers can be batched on a freight train to the corresponding train holon and the concerned order holons. Note that the staff holons only provide advice and that the basic holons are still responsible for taking the final decisions. This way, the concept of staff holons allows for the presence of centralized functionality in the architecture without introducing a hierarchical rigidity. This centralized functionality allows aiming for a good global performance, which is otherwise difficult to obtain as every holon tries to optimize its own (selfish) objective. To obtain its advice, a staff holon may rely on centralized scheduling algorithms, human input, artificial intelligence methods, etc. Next to scheduling advice, the staff holon can for instance provide advice about route planning or the balanced loading of a cargo ship. In case of scheduling advice, the various order holons will attempt to execute (the relevant part of) the provided schedule. They will deviate from the original schedule only if they find a significantly better solution or the provided advice appears to be (or has become) infeasible.

Interactions Between the Holons

As indicated in Figure 5.1, the various holons interact and share data with each other. The main interactions between the basic holons are briefly discussed hereafter. Remark that these interactions do not describe the dynamics of the holonic system described by the PROSA reference architecture; these dynamics are described by the DMAS coordination system described hereafter.

- *Product-order interaction:* The order holons interact with their corresponding product holon on how to correctly execute their task by using certain resources. After (virtual) execution of an operation, the order holon passes information about the resulting state and about next possible resources to the product holon. Based on this information, the product holon provides the order holon with all possible next operations. For instance, after loading a container onto a trailer, the corresponding order holon consults its product holon to know the following operation that should be executed. Usually, multiple options are available, for example, direct transportation to the final destination, transportation to an intermodal hub to be loaded onto a train or ship, etc.

- *Product-resource interaction:* Product and resource holons share process-related information. When generating a list of possible operations for an order holon, the product holon interacts with resource holons to know which operations the resources can perform. The other way around, the product holon provides the resource holon with technological aspects to correctly process an order, that is, the necessary process parameters to perform an operation. For instance, if refrigerated products have to be transferred between two trucks in a noncooled terminal, the product holon will indicate that this transfer should happen as fast as possible and impose a maximum transfer time.
- *Resource-order interaction:* The resource and order holons mainly interact to reserve operations on the resources. To this end, the resource holons provide the order holons with the results of virtually executed operations and reserve capacity when requested. Once an operation is started, the resource holon also informs the order holons about the execution result and progress. The desired coordination and control then emerges in a self-organizing way from the interactions between the various holons.

Concluding Remarks

During the development of the PROSA reference architecture, special attention was paid to *separation of concerns*. The three basic holons represent a separate concern of the underlying domain. An important characteristic of PROSA is for instance that resource allocation aspects and process-specific aspects are separated. The issue of technical feasibility (executing a task in a technically correct and validated manner) is addressed by product and resource holons. Product holons are knowledgeable concerning the capabilities of resources that are relevant to them, but they ignore resource capacity and availability aspects. The product holons inform the order holons about (all) technically correct manners to execute their task instance. Order and resource holons then address the issue of resource allocation.

Importantly, this separation of concerns safeguards and maximizes the potential for achieving critical user mass in holonic MES (HMES) and LES implementations. Here, the understanding that *membership of suitable autocatalytic sets* is crucial was put into practice.

Moreover, PROSA is a *structural decomposition*, mirroring its world-of-interest. This complies with the first principle of design for the unexpected. In particular, when a product holon mirrors what is possible, offering multiple options to order holons, it applies this principle when a conflict with a product holon equals a conflict with reality, in which no other known and validated option exists.

BIO-INSPIRED COORDINATION AND CONTROL IN HOLONIC EXECUTION SYSTEMS

Initially, PROSA implementations were heterarchical control systems. Their order holons would steer product carriers through a manufacturing system, visiting processing stations and having production steps performed as allowed by the product holons. The system would be myopic, lacking global optimization or coordination and operating much like automobiles in traffic.

An early exercise with a staff holon delivering scheduling advice was carried out. However, to have adequate collaboration, the order holons became dependent on specific properties of this advice and a reactive scheduler was needed to handle disturbances (Bongaerts et al., 2000). Hence, the research team started looking for a remedy for this myopia while remaining compliant with design for the unexpected.

The source of inspiration for this remedy was twofold. First, there was a heterarchical design in which workstations would "start emitting a signal" when they were about to become idle. This straightforward and simplistic idea was highly unsatisfactory. This signal arrives too late and fails to account for a product's journey needed to arrive at this station. Second, the behavior of food-foraging ants, called stigmergy (Grassé, 1959), revealed how to incorporate nonlocal information in a solution while employing only local reality-mirroring components, which fits D4U perfectly.

Food foraging ants execute a simple procedure:

- In absence of any signs in the environment, ants perform a randomized search for food.
- When an ant discovers a food source, it drops a smelling substance, called *pheromone*, on its way back to the nest while carrying some of the food. This pheromone trail evaporates if no other ant deposits fresh pheromone.
- When an ant senses a pheromone trail it will be urged by its instinct to follow this trail to the food source. When the ant finds the food source, it will return with food, while depositing pheromone itself. When the ant discovers that the food source is exhausted, it starts a randomized search for food and the trail disappears because of the evaporation.

This simple behavioral pattern results in an emergent behavior of the ant colony that is highly ordered and effective at foraging food while being robust against the uncertainty and the complexity of the environment.

An important capability of this type of stigmergy can be observed: global information about where to find food in a remote location is made available locally, indicating in which direction the ant must move to get to this food. Also, the complexity of the environment is handled in an elegant way by making the environment part of the solution (i.e., the complex shape of the pheromone trails), effectively shielding the ant colony solution from this complexity.

For the design and development of coordination and control systems, based on stigmergy, the following principles are recognized:

- Make the environment part of the solution to handle a complex environment without being exposed to its complexity. This complies with the essential modeling approach of object-oriented design.
- Place relevant information (pheromones) as signs in this environment ensuring that locally available data informs about remote system properties, supporting systemwide coordination.
- Limit the lifetime of this information (evaporation) and refresh the information as long as it remains valid. This allows the system to cope with changes and disturbances.

The combination of these sources of inspiration resulted, ultimately, in the *architectural patterns* that are discussed next. The signal became the local schedule of the resource holon. The deposition of pheromones was translated into order holons reserving time slots in these local schedules. The ability to answer what-if questions allowed "ant agents" to travel virtually and execute virtually what an order holon might "do for real." Details follow in the next Section on the Delegate Multiagent System.

The research translated this food foraging in ant colonies into a solution that remedies shortsightedness in decentralized coordination and control systems. The solution makes nonlocal information – using local reality-mirroring software components – locally available. Note that "nonlocal" is to be understood both in the geographical/spatial sense and in the temporal sense. It is about something both elsewhere and in the future. This part of the research was termed to be "predicting the unexpected" when presented to laypersons, unfamiliar with the application domain. This caused the adaptation of "design for the unexpected" as the nom de guerre for the research overall.

Delegate Multiagent System (DMAS)

This section belongs, as it immediately presents the generic and broadly applicable concept, in the next chapter. However, the DMAS is introduced here to avoid duplication and confusion over terminology. Consequently, the discussion in this chapter covers the specific DMAS designs used in the HMES research prototypes. The broader applicability is addressed in Chapter 6.

The *DMAS* is an architectural pattern. An architectural pattern is a concept that solves and delineates some essential cohesive elements of a software architecture. Typically, it comprises a description of software components and their interactions, together with a set of constraints on these components and interactions that define a set of architectures that satisfy them.

The DMAS pattern allows an agent – that is, a holon in the present discussion – to delegate a responsibility to a swarm of *lightweight agents*. These lightweight agents perform particular activities to support the issuing holon in fulfilling its functions. A holon can simultaneously delegate multiple responsibilities, applying the DMAS pattern for each of them. The holon may also use a combination of DMASs to handle just a single responsibility.

Figure 5.4 shows a module decomposition view of the architectural pattern. There are three modules: the environment, the agent, and the ant (Verstraete et al., 2008). The depends-on relationships between the modules are refined in the sense that the agent and ant module share data with the environment module, whereas the relationship between the agent and ant module is a "creation" relationship.

These lightweight agents are called *ant agents* or simply *ants*, after their biological source of inspiration. They are lightweight in the sense

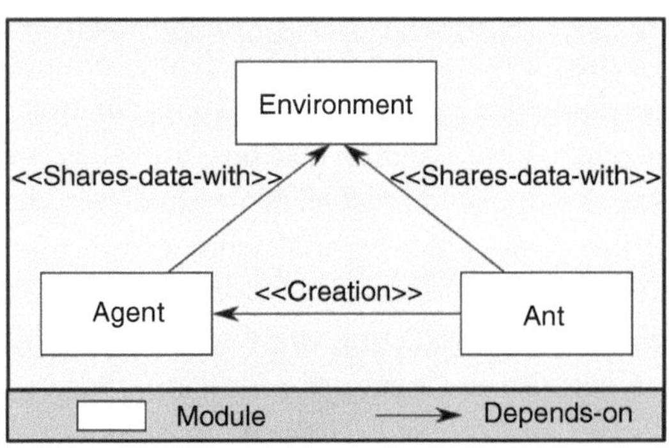

Figure 5.4 ***DMAS: Module Decomposition View.***

that each ant may only perform a bounded computational effort within its bounded lifetime and has a bounded footprint (memory). They are responsible for executing a task that serves a responsibility of the issuing agent/holon.

Each ant is created and initialized by its issuing holon. It (virtually) travels autonomously through the (virtual) environment. The ants start from a location selected by their issuing holon. Typically, this location is where this issuing holon resides (virtually), for example, the location of a product carrier. But, an issuing holon may create ant agents at a location from where finished products are shipped to their customer. From there, the ants travel (virtually) in opposite directions, typically toward the location of the issuing holon. Ants may even (virtually) traverse their journey twice, collecting information first and depositing information (digital pheromones) during the return journey.

Corresponding to the description used before, the *environment* is a software representation of the world-of-interest. To support navigation of the ants, resource holons know their neighbors (note that this is local information). This effectively provides a directed graph, possibly augmented by relevant information (e.g., maximum height), allowing ants to discover their world-of-interest starting from their initial location. Note that the evaporate-and-refresh mechanisms – copied and translated from the real-world ant colony behavior – ensures that reconfigurations and other changes will be mastered by the DMAS in an HMES or LES.

A holon delegating a responsibility to a swarm of ants is responsible for maintaining the population size and the diversity of this swarm. It chooses the creation frequency and initialization for every ant type. The individual ants are not aware of these swarm properties. The holons observe and interpret the (digital) pheromones in the environment and adapt their behavior accordingly.

Three types of DMASs are distinguished in the research prototypes: *feasibility, exploring, and intention DMAS* (Hadeli et al., 2004).

Feasibility Ants

A resource holon delegates part of its "information providing" responsibility to a swarm of so-called *feasibility ants*. These ants make global feasibility information (about the capabilities of the resource) locally available for the other holons. They put a kind of digital signposts on the blackboards of resource holons. They enable order holons to decide locally which routing options are available to them.

Feasibility ants are created by resource holons corresponding to an end point of the resource graph. The ants start at these nodes and traverse the graph upstream, that is, in opposite direction of the orders. During their trip, they collect information on the capabilities (e.g., type of supported operations) of the visited resources. This information is also deposited on the local blackboards of the resource holons and merged with the already available information from other feasibility ants. In this way, every node contains information about which capabilities are reachable via its connected nodes (somewhat similar to routing tables in computer networks). This information permits order holons (or their corresponding product holons) to determine which part of their process plan can be executed downstream. This activity is performed at a regular frequency such that changes (both in resource capabilities and topology) become quickly visible throughout the system.

For instance, for the resource graph shown in Figure 5.5, feasibility ants are created by the resource holons corresponding to D1 and D2. They travel upstream toward O1 and collect information about the provided (transportation) activities. The ants deposit this information at the nodes they encounter. In this way, the information at O1 will for instance indicate that D1 can only be reached via truck T1.

Exploring Ants

Every order agent generates explorer ant agents at a given frequency. These explorer agents are scouts each of which searches for an attractive route through the underlying production system, that is, to accomplish the given

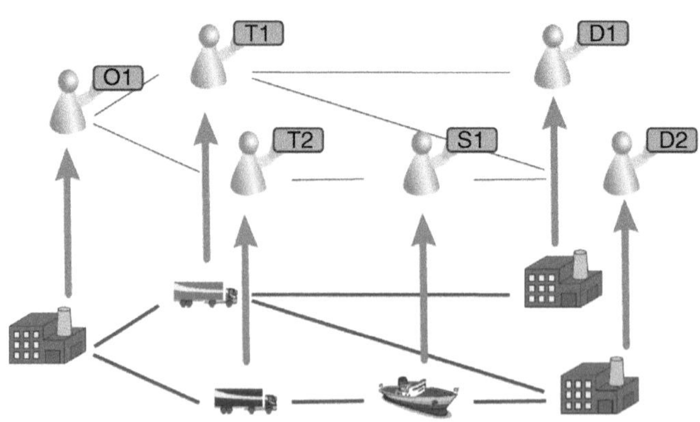

Figure 5.5 ***Example of a Resource Graph.***

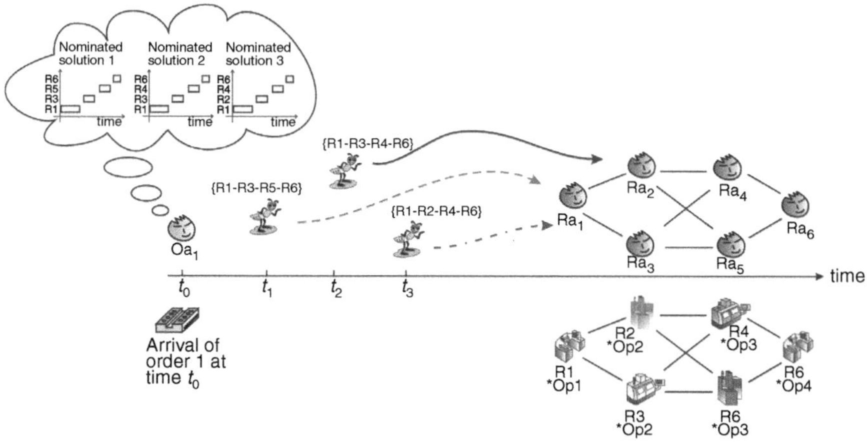

Figure 5.6 ***Ant Agents Explore Possible Routes.***

task (Figure 5.6). Depending on the performance criterion, these explorer agents search forward from the current state of the task onward (e.g., lead time minimization) or backward from the final delivery point (e.g., due date accuracy). Note that different order agents can have different performance concerns; rush orders, normal orders, low-priority orders, and maintenance orders have different objectives. The objective of a given order may even suddenly change (e.g., when a work piece gets damaged and needs a speedy replacement).

These scouts use the same method as the order agent, managing the actual execution of the task, to ensure that a proper sequence of processing steps gets executed, but virtually. The feasibility concern is handled by the feasibility ant agents. As explained, these ants deposit information on the information spaces (blackboards) attached to entries and exits of the resources that allows the product agents to discern valid and invalid routings locally. This information also evaporates and is refreshed to account for changes in the production system.

The search strategy employed by the explorer agents is a plug-in of the control system. Not every explorer ant uses the same strategy. Typically, some percentage looks for the promising routes whereas other ant agents look for solutions that aim to avoid critical resources. The key point is that the emergent forecasting does not rely on which strategy is employed by these scouting agents.

During its exploration journey, ant agents delegate the information processing to the product holon and resource holons. Their product holon provides the set of legal routing options that are open to the scout at each routing point. It makes sure that the product recipe is obeyed. The resource agents provide the necessary performance estimates.

Consider an explorer ant moving forward from the current position of a work piece on a conveyor belt. The scout queries the conveyor belt holon about the estimated remaining traveling time on the belt. The explorer ant then virtually travels to the exit of the belt and adjust its internal clock indicating the expected arrival time. At this point, the scouting ant selects a legal entry connected to the exit of the conveyor belt and continues its virtual journey.

On arrival at a processing unit, the product holon indicates the possible processing steps and the explorer ant selects a processing step. The explorer agent queries the resource holon of the processing unit about estimated queuing and processing times as well as processing results. In this manner, the virtual journey continues until the final delivery point is reached. Note that the resource holons will virtually execute their own execution strategies. The emergent forecasting mechanism does not require any specific strategy from these resources. Again, a strategy is a plug-in for the control system.

When an explorer ant agent has virtually executed the task, it reports back to the order holon. The report includes the journey and the performance estimates of that journey. Based on the results of its exploring ants, the order holon keeps a set of candidate routes. These candidates get refreshed regularly, either explicitly by specialized exploring ants that simply follow a given route, or by ensuring that the normal exploring ants will rediscover these currently attractive candidates with a high probability. The set of candidate routes is selected based on the performance estimates and on their complementary nature (i.e., limit the number of candidates that have very similar routings). The candidates that have become too old are eliminated from the set of candidates by evaporation. Basically, the exploring ants implement a distributed heuristic search for good production schedules and adapt the solution continuously to account for changes and disturbances. The optimization heuristic itself is not the focus of our considerations here, *as its selection and configuration inherently cannot be designed for the unexpected.*

Intention Ants

The above-mentioned exploration requires the resource holons to possess an adequate estimate of their future workload. The order holons generate intention ant agents, at a given frequency, to serve this purpose. When

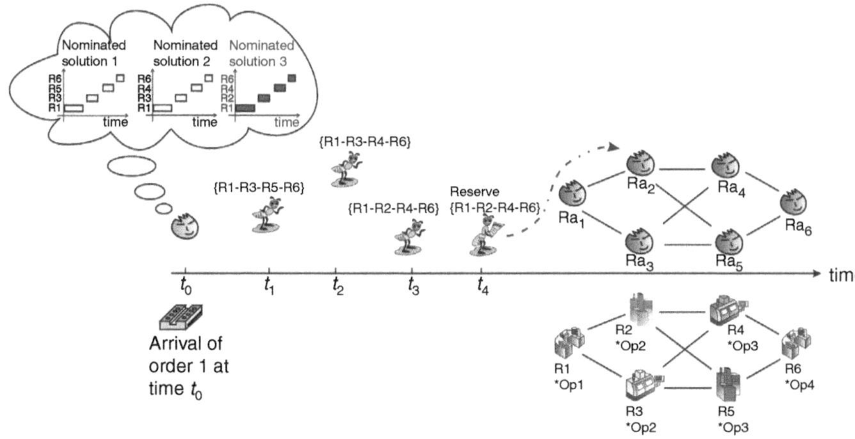

Figure 5.7 ***Ant Agent Propagates the Order Intentions.***

a suitable set of candidate solutions has been constructed (see Section on Exploring Ants) and the estimated starting time for the processing of the product instance(s) approaches, the order holon selects one of the candidate solutions to become its intention. Then, the order holon generates intention ant agents to notify the holons of the affected resources of its intentions (Figure 5.7).

The intention notification service operates as follows:

- The intention ants virtually execute the routing and processing of their selected candidate solution. On their virtual journey, the intention ants acquire travel, queuing, and processing times from the resource holons on their path. Any changes, which occurred since the exploration, immediately become visible when these resource holons provide the information.
- In contrast to the exploring ants, intention ants inform the resource holon that their order holon is likely to visit them at the estimated time. In this way, intention agents make a (evaporating) booking on the resource, and the resource holon adapts its load forecast (local schedule of the resource) to account for the visit of which it is informed by an intention ant. In other words, the intention ants enable an emergent forecasting of the resource utilization. As a consequence, resource holons are able to predict performance more accurately to their visitors, the exploring and the intention ants.
- The intention information at the resource holon evaporates. Order holons must create intention agents at a refresh frequency that is sufficiently high to maintain their bookings at the resources.

- While refreshing, the order holons observe the evolution of the expected performance of their current intentions (through intention ants' reporting their estimated performance). This performance estimate is compared to the estimates of the candidate solutions that are found and refreshed by the exploring agents.
- When the estimated performance of the current intentions drops significantly below the estimated performance of some candidate solutions, an order holon may change its intentions.
- When the product instance(s) of an order holon reaches the point where a decision needs to be executed, the order holon triggers the action in the underlying system in accordance with the intentions. If an event occurred that makes this impossible or highly unattractive, the order holon delays the action shortly such that the above procedure may find a (better) solution.

Short-term Forecasting – Predicting the Unexpected

The combination of exploring and intention DMAS provides a view on the short-term future of the system, which is based on an estimation constructed through a decentralized virtual execution (i.e., a simulation embedded in the HMES). Both resource and order holons have short-term forecasts about their predicted execution. The order holons know the expected routings and resource allocations for their orders. The resource holons know the predicted loads for the corresponding resources.

The resource holons receive the necessary information to calculate a short-term forecast of their utilization via the intention DMAS. Based on these forecasts (and their what-if functionality), they are able to give accurate answers to the queries from the exploring ants. This in turn allows the order holons to have a precise view on their short-term future. Note that the order holons create exploring and intention ants at regular time intervals, even after they have selected an intention. This allows them to react to disturbances and new opportunities and keeps the short-term forecasts up-to-date.

All short-term forecasts together can be seen as a "dynamic schedule." Figure 5.8 shows an example of such a schedule. This way, these forecasts provide visibility of future actions, which is recognized as a valuable property. This visibility allows for instance to identify potential capacity conflicts, permitting management to take action on time to avoid them. It also facilitates operators to see the bigger picture and to anticipate the impact of their decisions.

Moreover, in the context of manufacturing, creating visibility on the shop floor is considered as one of the main goals of an MES. These

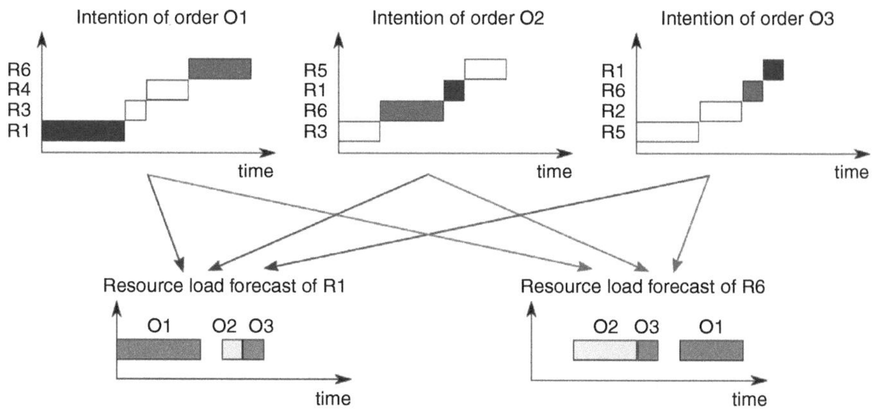

Figure 5.8 ***Example of Generated Short-Term Forecasts.***

short-term forecasts can be used by the order and resource holons to take better (informed) decisions. For instance, as the order holon has an accurate view on its intention, it can make a well-considered decision whether or not to switch to one of its alternative solutions. The resource holons know their expected loads and can for instance decide to process a rush order or not, based on how many reservations will be affected. As another example, the resource holon corresponding to a truck can, based on its expected load, decide when it is a good moment for maintenance or refueling.

For these forecasts to be valid and reliable, the order agents cannot continuously change their intentions, making the system overly nervous. This nervousness issue is addressed in the next section. Finally, in the Section, Cooperation of HMES with Planning Systems, the cooperation of the HMES with external planners and schedulers is discussed.

SOCIALLY ACCEPTABLE BEHAVIORS FOR DMAS

The DMAS pattern makes information available in locations that are not colocated with its *source of truth* (the word "location" is used in a virtual sense). For instance, order holons are aware of the available resource capacities and capabilities through information collected by a suitable DMAS. Moreover, holonic execution systems use this DMAS technology to generate short-term forecasts, both for order routings and resource utilization. In other words, nonlocal information is provided both in the spatial (e.g., inform an order holon about a distant resource) and in the temporal sense (e.g., inform an order holon about a future state of this distant resource).

However, there is no such thing as a free lunch. The information collected by a DMAS is only a snapshot. It is a copy and there is no guarantee that it will remain correct. An intention ant may have "refreshed" its order holon intentions when, one second later, because of a machine breakdown, this intention becomes invalid. In other words, there is no "lock" on the source of truth to guarantee that information remains valid. Note that reality (the sun is shining) cannot be locked (two minutes later, it rains), and therefore locking is not an option.

This situation has implications for our ability to generate reliable and usable short-term forecasts. Indeed, our predictions are the results of a virtual execution of the order holon intentions. When these intentions change, the generated predictions may change. As order holons select their intention in function of expected performance, which depends on the predictions, one can imagine scenarios in which the predictions become inaccurate, useless, and even harmful.

Indeed, order holons observe – through their DMAS – the predicted system states and select their intention (e.g., congestion-free highway to Brussels). These intentions are used to generate/adapt the predictions (e.g., too many order holons selected this highway, and now the prediction indicates serious congestion). In the new prediction, other solutions – discovered by the exploring DMAS – perform better and the holons change intention (e.g., the smaller roads are predicted to be congestion-free). Too many order holons made the same or a similar switch of intentions, and the prediction changes again. If the system lacks a suitable dampening mechanism, the emergent short-term forecasting will have a serious problem.

To avoid, order holons have to behave in a socially acceptable manner. Today, humans already have such behavior. When people make an appointment or have a working arrangement (e.g., which lecturer will examine which students at what time and location), they do not change this without good reason and they minimize negative impact to the remainder of society. Research on designing such behavior in order holons is discussed later.

How Order Holons Change Their Intention

Order holons have an exploring DMAS that continuously discovers candidate solutions. This DMAS also refreshes top-performing candidate solutions. At some point, the order holons have selected such a candidate solution to become their intention, and their intention DMAS regularly propagates their intention by means of virtual execution.

When the estimated performance of the top-performing candidate solution is better than the performance estimate for the intention (based on the most recent intention refresh), the order holon probably wants to switch and select this top-performing candidate as its new intention. This switching has been the subject of research about incorporating a sound resistance to change intentions. However, such resistance must not be too strong as this would deny the order holons to react to changes and disturbances.

Mechanism to Dampen Intention Switching in Order Holons

The research designed, implemented, and evaluated a combination of three dampening mechanisms:

1. *Thresholding*. The order holon only decides to change intention when the expected performance improvement is above a given threshold. At least, this may and must limit changing of intentions to "only for improvements well above the noise level of the estimation mechanisms."
2. *Cool-off period*. When an order holon effectively changes intention, its threshold is temporarily raised (significantly). This gives other order holons the opportunity to adapt while this order holon sticks to its plans.
3. *Randomized switching*. Recall that the applicability of our holonic execution systems favors real-world systems that are much slower than their virtual execution. Therefore, there will be many refresh cycles (of intentions) and explorations while in reality nothing much happens or changes. In other words, order holons may delay switching intentions in terms of refresh cycles without a negative effect in the real world; the switch will be fast enough to have its effect in reality. Randomized switching capitalizes on this. When an order holon wants to switch (e.g., the estimated performance improvement is above the threshold), there is only a chance that this will happen during the next refresh cycle. Based on a genuine random number, a lottery mechanism ensures that only a fraction of the order holons, wanting to change intentions, switch at the first opportunity. An antistarvation mechanism in this lottery ensures an upper bound on the number of refresh cycles an order holon may have to wait.

Switching Intentions in Case of Major Disturbances

The above dampening mechanisms have been designed with the handling of major disturbances in mind. When an important piece of equipment breaks down or a large rush order enters the factory, a significant number of order holons will discover that the performance estimate for their intentions has deteriorated significantly. After refreshing their top candidates, they are likely to be wanting to switch based on the illusion of free capacity on the alternative resources.

Here, the randomized switching prevents a stampede to such alternative resources. Only a smaller number of order holons will switch and the increased use of these alternatives will result in lower performance estimates when the other holons refresh their top candidates. In the meantime, the estimated downtime for the equipment that broke down often will become more accurate (i.e., less conservative). Thus, the randomized switching ensures a steady stream of order holons switching to alternative resources until "waiting for the repair" goes below the threshold (for the order holons that did not switch yet).

Experiments, on relatively small problems in simulation, revealed a relatively large sweet spot. Both "imposing a small amount of dampening" and "only allowing to switch when there was really good cause to do so" have been effective. There was no indication that the tuning of this combination of dampening mechanisms was challenging. The main conclusion of these modest experiments was that it is important to have the dampening mechanism in place (Hadeli., 2006).

Evidently, the ability to execute many refresh cycles and to spread the changing of intentions over a lot of cycles will render dampening and its tuning easy-to-do-well. If randomized switching needs to be performed in a low number of cycles, tuning is likely to become an issue. Indeed, no order holon will have the opportunity to observe what is happening at a system level before switching. In such cases, the tuning has to address this for the order holons, which is likely to make tuning a case-dependent task (that needs to be redone when conditions change).

Concluding Remarks

The main contribution of our research was to identify the need for dampening of the intention switching by order holons. At the level of the society of holons, this prevents a harmful system nervousness that may render the emergent forecasting ineffective. The research has only investigated the simplest manners of dampening. Challenges for future research include the following:

- Changing intentions in small teams. Humans apply this quite often. For instance, they change intentions in pairs when they switch slots.
- Changing intentions in open organizations. The above research looked into problems with a single owner/organization: there is a boss to impose how to behave. When looking into problems with multiple owners/

organizations/individuals that strive for their own interests, especially, the randomized switching faces additional issues. Indeed, it is unlikely that participants will delay switching unless a suitably authorized arbitration service imposes this altruistic behavior. Moreover, there may be a legal requirement to treat all "order holons" on an equal footing. This implies that this arbitration service has to ensure that a good deed (which it imposes) will be properly rewarded (or at least remains unpunished). This remains an unaddressed and unresolved challenge.

- Detecting and measuring system nervousness to be able to adapt dampening in function of the situation at hand equally remains uninvestigated. Likewise, identifying where and when in a system dampening and nervousness are important remains an open issue.

However, the positive message from modest experiments is that, especially where high refresh rates are possible, the dampening of intention switching to control system nervousness – to protect the validity/usability of emergent forecasting – is first of all a matter of providing suitable mechanisms where the tuning of these mechanism proved to be undemanding and straightforward. The reader is referred to Hadeli (2006) to find worked-out examples.

COOPERATION OF HMES WITH PLANNING SYSTEMS

Production in a factory is generally organized according to some kind of planning scheme that is created before production starts. From an MES perspective, such planning hardly ever provides all relevant details. While it normally specifies which jobs need to be performed on which machines, in what time period, it does not specify details like which transportation unit to use for carrying parts from and to machines, where to store parts, which tools to use, etc. A planning is released on a regular basis, weekly, daily, or every shift or hour.

In practice, production activities deviate from this planning within the proverbial first minutes. The production floor continues to use release dates and due dates from the planning but otherwise manages the activities on the shop floor autonomously. In fact, the planning mainly serves for the remainder of the organization, not the production itself, to stay realistic about what production may and will achieve (and what would be unreasonable to expect).

Admittedly, the above reflects the gap between industrial practice and academic theory, where the latter certainly is more advanced and this gap is closing rather rapidly. However, further developments in planning technology will not solve the problem. The planning/scheduling problem is

NP-hard or worse, and a planning/scheduling algorithm must be computationally efficient if it is to be used in practice. If such an efficient algorithm were to solve this combinatorial problem without making (partially arbitrary) choices, discarding options, and so on, it would (to provide an estimate how unlikely this is) break most of the currently used encryption in computer networks.

Therefore, it is fairly safe to conclude that the planning functionality – when made operational in a manufacturing system – cannot be designed for the unexpected. More precisely, the planning/scheduling algorithm itself is designed for the unexpected: it just exists and can be considered as an element in the world-of-interest (e.g., Karmarkar's algorithm for linear programming). It is the selection of a specific algorithm and the manners in which it is connected and configured to the manufacturing systems that are forced to rely on assumptions, rendering this a design for the expected.

However, manufacturing organizations have to reconcile two objectives concerning shop floor operations:

- Optimization of production performance relative to the management goals (e.g., reduce costs, satisfy customer demand). Today's planning systems have been conceived and constructed to address this concern.
- Robustness and thoroughness, or the realization of the production objectives derived from management goals, while accounting for all the relevant details and while handling uncertainties and unforeseeable disturbances.

Manufacturing execution systems and especially HMESs have been developed to address this latter concern. However, a pure self-organizing HMES cannot solve the problem by itself. Therefore, cooperation between an HMES and a planning system is indicated such that good performance, assured by the planner, is reconciled with robust execution, taken care of by the HMES. This is achieved through *schedule execution* by the HMES.

Schedule Execution

Schedule execution is the process of taking on-line resource allocation decisions, based on an existing schedule but also considering the actual state of the resources and orders on the shop floor.

Handling a production order in the self-organizing HMES implementation is achieved using the exploring and intention ant mechanisms explained in the Section "Bio-Inspired Coordination and Control in Holonic Execution Systems." A fast and frequent virtual execution of the envisaged production activities, by these mechanisms in a virtual world, allows to quickly detect unexpected events and to come up with alternative solutions enabling to remedy problems and grasp opportunities.

However, every order holon or resource holon considers optimality from its own – often selfish – perspective in the basic HMES implementations. Therefore, research has developed and assessed versions of the exploration and intention mechanisms that take an existing schedule into account. Note that this research had no ambition to contribute to planning/scheduling knowledge; only the cooperation was addressed.

To make the best trade-off between sticking as close as possible to the original schedule or to resort to a new schedule, two processes are introduced: *last-mile planning* and *online optimization*. The last mile-planning process aims to keep execution as close as possible to the planning. The online optimization process tries to find a new solution from scratch, while the production is running. Both processes run concurrently and take resource allocation decisions. The schedule execution process balances both processes. Summarizing, schedule execution is the combination of last-mile planning and on-line optimization.

How can last-mile planning and on-line optimization be combined?

Schedule execution requires the order holons to anticipate potential disturbances, by executing an adaptive search. Therefore, order holons send out two types of exploring ants:

1. *Type-1 exploring ants.* They ignore the schedule and search for alternative solutions in the search space in a randomized manner. While these ants cannot cover the huge search space, they might still find interesting alternative solutions that their order holon can use in case of disturbances.
2. *Type-2 exploring ants.* They cover the search space "around" the schedule. This type of ants makes sure the required auxiliary operations are selected while executing the planning. Indeed, these ants have to fill in the missing details in the schedule and also have mechanisms to approximate the schedule if it is impossible to follow it completely (e.g., when a resource becomes available shortly after the time specified in the schedule).

The intention selection mechanism in the order holon will favor solutions from type-2 exploring ants, except when the schedule information deviates significantly from the performance and behavior predictions, which originate from the virtual execution by the ant. In other words, if exploration reveals that the schedule is feasible, it will be followed. If exploration by type-2 ants reveals the schedule to be unfeasible or forced to deviate considerably, the order holons assumes the schedule to be invalid and have all candidate solutions compete as equals.

Resource holons equally favor visits according to schedule. When intention ants request the resource holon to reserve capacity, time slots according to the schedule have the priority over allocations/reservations deviating from the planning. This avoids that disturbed orders cause a cascade involving nondisturbed orders.

The relative amount of each type of ants can be established in two ways:

1. In a very dynamic environment, the schedule may be outdated often and rapidly. Therefore, the reactivity of the order holons to new opportunities may be more important than the execution of the schedule. However, parts of the schedule may still be useful. In this case, the most appropriate relative amount of exploring ants can be determined by simulating a number of expected disturbances and investigating their effect on performance.
2. In more predictable environments, executing the schedule may be more important. Unexpected events occur infrequently. In that case, the percentage of exploring ants that cover the search space "around" the schedule determines how fast the MES reacts to variations in executing the schedule. In addition, the quality and completeness of the schedule determines how much exploration around the schedule is optimal. If the exploring ants have little to add, a small number will suffice.

Experimental Verification

Manufacturing Example

The cooperation of an HMES with a general-purpose planning system has been tested in a few simple case studies, sometimes with surprising, counterintuitive results.

The first example consisted of a flexible manufacturing system where a number of workstations (machining centers, cleaning station) were connected by an automated storage and retrieval system (AS/RS) (Figure 5.9). A crane delivers raw materials to these workstations and collects finished products from them. The operations of both the workstations and the AS/RS are highly interconnected.

An offline planning system, LEKIN, performs a global make-span optimization, whereas the HMES performs a decentralized local optimization (of average lead time). LEKIN is a general-purpose academic planning tool. It ignores some of the aspects of the production system, such as transportation and warehouse management.

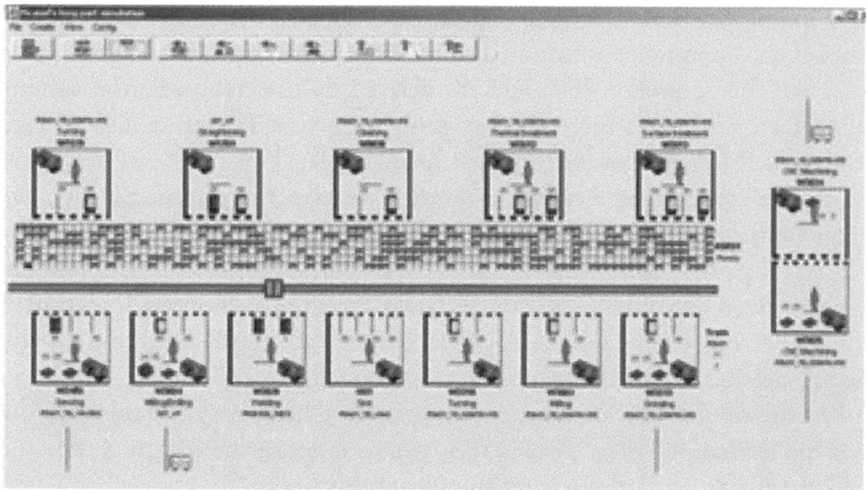

Figure 5.9 ***Flexible Manufacturing System Layout (Screen Shot).***

The experiments considered three different process plans, two of which have alternatives. They investigated the influence of the (in)accuracy of the LEKIN planner, of the workload (8 or 15 orders), and of the level of guidance by the planner (self-organizing with 100% type-1 exploring ants; schedule execution with 25% type-1 exploring ants; last-mile planning with 100% type-2 exploring ants).

Table 5.1 summarizes some relevant experimental results. The most striking result is that in normal operation (no breakdowns), the average lead-time is smaller for the system working in the self-organizing mode than when the planning is taken into account. This is caused by the scheduling rule used in the AS/RS. The AS/RS cannot serve all orders at once. Depending on the workload of the manufacturing system, an order may therefore suffer a considerable delay, sometimes of more than one hour.

Table 5.1 Influence average lead-time of different levels of interaction between HMES and planner

		Normal operation			Machine breakdown		
8 orders		SO	SE	LMP	SO	SE	LMP
	Correct planner	1660	1900	1894	1938	2010	2653
	Faulty planner	1753	2144	2058	2127	2168	2831
15 orders	Correct planner	2458	2677	2597	3057	2934	3543
	Faulty planner	2554	2830	2814	3941	3170	5086

SO: self-organizing mode; SE: schedule execution (25% type-1); LMP: last-mile planning

Depending on the position the order has, the delay may be more or less important. A sequence is imposed on all orders and the orders cannot deviate from this sequence. If orders are delayed by the transportation system, this effect propagates through the whole sequence. Therefore, the average lead-time is larger than for the self-organizing level.

At the 75% guided level, all orders deviate from the planning if their average performance deviates from the planning above the predefined threshold. As long as the delay imposed by the transportation system is smaller than the threshold, the orders will not deviate. Therefore, the average lead-time is still higher. However, because all orders that deviate above the threshold will search another solution, the variation is reduced.

As the results indicate, an incorrect planning has a negative effect on the average lead-time when the HMES tries to execute the planning. For the self-organizing level, there is no significant difference.

To examine the responsiveness of the HMES, the effect of a machine breakdown is investigated. The basic scenario remains the same. The first time an order is executed on one workstation, however, it breaks down for about 16 h. During this time, the order cannot be removed from the workstation. Once the workstation is repaired, the order can continue its execution. The experiments investigate the interaction between the machine breakdown and the other factors, shown in Table 5.1. The experiments confirm that the schedule execution approach improves the responsiveness of the HMES toward breakdowns. Both the variation and the average flow time suffer less from a breakdown compared to an approach where the HMES only executes the planning or uses no planning at all.

This responsiveness is achieved while maintaining a good separation of concerns. On the one hand, the planning system – in this case the LEKIN scheduling system – can be used unmodified. The HMES adds responsiveness to the functionality the planning system already offers. On the other hand, the HMES compensates for simplifications in the planning system. For this study, the scheduling system ignores transportation and warehouse management and cannot deal with parallel machines with different execution times. The introduction of the planning system in itself did not bring a general improvement in average performance. This is mainly due to the variation the transportation subsystem introduces. When following the planning, this effect is increased. Therefore, in general, an assessment of the variation in the execution system before introducing a planning technique may be appropriate. For this case, alternative scheduling rules at the level of the transportation system may improve the effect of the planning on the execution system.

Finally, note that these experiments remain very limited in scope. They were performed before the introduction and use of Erlang/OTP for the prototype developments. In addition, the schedule-generating system (LEKIN) was elaborated for this experiment only. There certainly remains an unaddressed gap to be addressed in future research in which this cooperation is investigated more thoroughly, especially the codesign of the scheduling and the last-mile planning by the type-2 exploring ants. In fact, the team developing the scheduling part is also the natural provider of plug-ins for the type-2 ants, defining what it means to stay close to a given schedule.

Logistics Example

The following simple logistics example further shows the potential of an HLES receiving advice from external schedulers. Figure 5.10 shows the resource graph for this application. One cross-dock is considered and orders have to be transported from their origin to their destination via this cross-dock. Each order consists of one pallet that has to be transported. Several trucks are available at the cross-dock, which are controlled by the cross-dock manager. These trucks are responsible for pickup of the orders at their origin and for delivering them at their destination. The trucks do not perform direct transportation from an origin to a destination, and each truck can transport up to 33 pallets at a time. Inside the cross-dock, one or more forklift trucks are available to unload the arriving goods, to transport

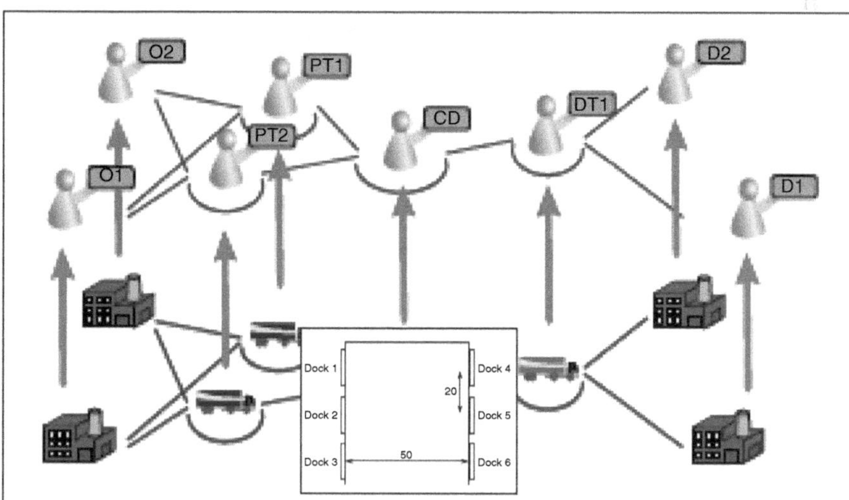

Figure 5.10 ***Layout of the Cross-Docking Application.***

these goods, and to load them into the correct outbound truck. The orders can also be transported to a temporary storage area. It is assumed that equipment and personnel is always available, that is, breaks or shifts are not taken into account. Another simplification is that the sequence in which a truck can be loaded or unloaded is not considered. If a delivery truck has finished its operations, it stays at its current position and does not return to the cross-dock (although the vehicle routing scheduling system [VRSS] assumes that all trucks make complete tours starting and ending at the cross-dock).

The considered cross-dock (CD) has six dock doors. Inside the cross-dock, two forklift trucks (with driver) are available to process the orders. Temporary storage of these orders is possible. The orders have to be picked up at one of two possible origins (O1 and O2) and delivered to one of two destinations (D1 and D2). The expected travel times between the cross-dock, origins, and destinations are known. There are three trucks available at the cross-dock to perform the transportation, two for freight pickup at their origins (PT1 and PT2), and one for delivery to their destinations (DT1).

In Experiment 1, eight orders have to be transported. Control is by an HLES according to the PROSA/DMAS architecture, without advice from a staff holon. Table 5.2 shows the performance measures of two simulation runs. In the first run, both pickup trucks visit another origin (PT1 visits O2, PT2 visits O1). PT2 picks up all orders in one time (orders 2, 3, 4, and 8), whereas PT1 only picks up order 6. PT2 then picks up the remaining orders at O2 (orders 1, 5, and 7) after it has delivered the orders from O1 at the cross-dock. Shortly after the (first) arrival of both pickup trucks, the delivery truck DT1 leaves the cross-dock with all orders for destination D2 (orders 2, 3, 4, and 6). After delivery, DT1 returns to the cross-dock to load the remaining orders (orders 1, 5, 7, and 8) and to bring them to D2.

Experiment 2 has the same setup as Experiment 1 but a staff holon gives an initial advice to the order holons (from a VRSS) and the cross-dock holon (from the truck scheduling system [TSS]). The performance measures are shown in Table 5.2. All orders at O1 are picked up in one go

Table 5.2 Performance measures of an HLES without scheduler advice (Experiment 1), with scheduler advice (Experiment 2), and with a breakdown (Experiment 3)

	Average tardiness (min)	Average flow time (min)	Make-span (min)	Total travel distance (km)
Experiment 1	298	1054	1950	3403
Experiment 2	183	954	1710	2507
Experiment 3	337	1150	1910	2507

by PT1, and all orders at O2 by PT2. When both trucks have arrived at the cross-dock, the eight orders are transferred to DT1, and this truck makes a tour via D2 to D1 to deliver all orders. Experiment 2 not only indicates that the HLES can cooperate with external scheduling systems but also that this cooperation improves the performance. Compared to experiment 1, all performance measures have ameliorated (see Table 5.2). This can be explained by a good batching of the orders, as advised by the staff holon. The performance measures also have improved compared to experiment 1.

In Experiment 3, the setup is the same as Experiment 1, but a breakdown occurs. One of the pickup trucks breaks down when it is transporting orders to the cross-dock. The control mode used is such that the staff holon recalculates its advice based on the current situation (i.e., truck positions). PT2 breaks down at time t = 772.0, when it is moving the orders originating from O1 to the cross-dock. The truck is repaired at t = 951.2 and continues its trip to the cross-dock. At that time, PT1 has already arrived at the cross-dock. The loading of delivery truck DT1 only starts once PT2 has arrived at the cross-dock and all orders are delivered in one tour (via D2 to D1). The corresponding performance measures are also shown in Table 5.2.

Other convincing experiments are described in detail in Van Belle (2013).

The experiments confirm that the HLES is able to cooperate with external scheduling algorithms and that this cooperation improves the performance. Moreover, if some aspects are not taken into account by the scheduling algorithms, the HLES is able to deal with these aspects. This approach allows for a cooperation scheme in which the staff holon gives advice to the orders and resources when a larger disturbance (e.g., breakdown) occurs, and the order, resource, and product agents deal in a self-organizing way with the smaller deviations in between. The experiments also indicate that the holonic system provides visibility about the current and future resource and order states. This allows the responsible decision makers to intervene on time if necessary or desired.

CONCLUDING REMARKS

Based on the Design Principles and the Laws of the Artificial, expounded in Chapters 3 and 4, respectively, a Holonic Manufacturing Systems description and control framework has been developed, able to provide an answer to many challenges present-day and future manufacturing systems are exposed to: robustness, scalability, foresight, autonomy, social behavior.

This PROSA/Delegate-MAS framework has been the solid base on which extensions and generalizations could be built, enabling the application of the framework beyond the manufacturing world. This generalization is the subject of Chapter 6.

ABBREVIATIONS

ADACOR	Adaptive holonic control architecture
AS/RS	Automated storage and retrieval system
CIM	Computer integrated manufacturing
DMAS	Delegate multiagent system
Erlang	Programming language used to build massively scalable soft real-time systems
HMES	Holonic manufacturing execution system
HLES	Holonic logistic execution system
LES	Logistic execution system
LEKIN®	Educational scheduling tool
MES	Manufacturing execution system
OTP	Set of Erlang libraries and design principles providing middle-ware to develop these systems
PROSA	Product-resource-order-staff architecture
TSS	Truck scheduling system
VRSS	Vehicle routing scheduling system

REFERENCES

Babiceanu, R.F., Chen, F.F., 2006. Development and applications of holonic manufacturing systems: a survey. J. Intell. Manuf. 17 (1), 111–131.

Bongaerts, L., Monostori, L., McFarlane, D., Kádár, B., 2000. Hierarchy in distributed shop floor control. Comput. Ind. 43 (2), 123–137, Special issue on intelligent manufacturing systems.

Dilts, D.M., Boyd, N.P., Whorms, H.H., 1991. The evolution of control architectures for automated manufacturing systems. J. Manuf. Syst. 10 (1), 79–93.

Grassé, P.-P., 1959. La reconstruction du nid et les coordinations Inter- Individuelles chez Bellicositermes natalensis et Cubitermes sp. la théorie de la stigmergie: Essai d'interprétation du comportement des termites constructeurs. Insectes Sociaux 6 (1), 41–80.

Hadeli, Valckenaers, P., Kollingbaum, M., Van Brussel, H., 2004. Multiagent coordination and control using stigmergy. Comput. Ind. 53 (1), 75–96.

Hadeli. Bio-inspired multi-agent manufacturing control systems with social behaviour. PhD thesis. KU Leuven, 2006.

Koestler, A., 1967. The Ghost in the Machine. The Macmillan Company, Hutchinson, UK.

Van Belle, J. A holonic logistics execution system for cross-docking. PhD thesis. KU Leuven, 2013. ISBN 978-94-6018-749-0.

Van Brussel, H., Wyns, J., Valckenaers, P., Bongaerts, L., Peeters, P., 1998. Reference architecture for holonic manufacturing systems: PROSA. Comput. Ind. 37, 255–274.

Verstraete, P., Saint Germain, B., Valckenaers, P., Van Brussel, H., Van Belle, J., Hadeli, 2008. Engineering manufacturing control systems using PROSA and delegate MAS. Int. J. Agent-Oriented Softw. Eng. 2 (1), 62–89.

Wyns, J. Reference architecture for holonic manufacturing systems: the key to support evolution and reconfiguration. PhD thesis. KU Leuven, 1999. ISBN 90-5682-164-4

CHAPTER SIX

The ARTI Reference Architecture – PROSA Revisited

Paul Valckenaers*, Hendrik Van Brussel**
*Faculty of Engineering Technology, KU Leuven
**Faculty of Engineering Science, KU Leuven

This chapter presents consolidated research results and the inroads made into new application domains. It discusses the generically applicable results. The consolidated research results connect the insights from Chapter 2 to Chapter 4 to a methodological development approach, an improved reference architecture, architectural patterns, etc. These results account for needs that emerged from various application domains, other than the manufacturing domain, as well as requirements from challenging real-world manufacturing cases.

A key element in the consolidated research results is the Activity-Resource-Type-Instance (ARTI) reference architecture. The most compelling factor, urging us to put forward this product-resource-order-staff architecture (PROSA) refinement, was terminology. PROSA terminology is manufacturing-specific. Applying PROSA to neighboring application domains (e.g., logistics) proved doable but far from ideal. When entering more remote domains (e.g., health care), PROSA terminology became unsustainable: communication with domain specialists and practitioners is crucial for a successful introduction of any innovation, especially when it requires that the persons involved stretch or even leave their comfort zone. In other words, the need for a generically applicable terminology constituted sufficient grounds, on its own, to revisit PROSA and propose the ARTI reference architecture.

Apart from introducing a universally applicable terminology, the ARTI reference architecture introduces a refinement. ARTI splits the resource holons (in PROSA) into resource type holons and resource instance holons. This mirrors the split of responsibilities between the product and order holons, which become activity type holons and activity instance holons respectively. This constitutes a more orthogonal basis, with resource/activity on one axis and type/instance on the other axis. The advantage over PROSA is a better-suited design when resource instances are employed in multiple roles that are situation or time dependent. Nonetheless, every

Design for the Unexpected. http://dx.doi.org/10.1016/B978-0-12-803662-4.00006-0

ARTI implementation will be a PROSA implementation (using a different terminology). Conversely, some PROSA implementations will not be ARTI compliant (e.g., when using specialization or, in programming jargon, inheritance to combine a resource instance and its type).

The discussion of inroads into new application domains covers aspects that have not (yet) been incorporated into the consolidated research results. They do not invalidate these results but still require time and (human) resources to be investigated further and developed in full. Often, these explorations into novel application domains turn nice-to-have requirements (in earlier research) into must-have requirements (originating from a new domain).

The Sections "Software/System Development," "The ARTI Reference Architecture," and "The DMAS Architectural Pattern" address consolidated results, and the Sections "Challenges and Lessons Learned from Applications" and "Toward a Humane Mechatronic Society" discuss the inroads. But first, the relationship between design for integrate-ability and design for the unexpected is revisited.

INTEGRATED WITH REALITY IN THE D4U PREFERRED MANNER

As mentioned in the Introduction, the research initially focused on integrate-ability. Building on a theoretical model, disclosing mechanisms preventing effective and successful integration, the research results transpired into design for the unexpected. Indeed, designing components and subsystems for integration, while assuming that integration requirements inherently are unpredictable, simply cannot be satisfied by anything less than coping with the unexpected.

Looking at integration in practice, a leader-followers pattern can be observed. Many small systems (followers) adapt to a big dominant system (the leader). Reality occupies this leader role when developing execution systems by definition. In other words, execution systems are to be designed for integration with their world-of-interest. Also note that the present state of the art provides no (peaceful) answers concerning leader–leader integration.

Design for the unexpected, and more precisely the research results discussed in this chapter as well as the previous one, exploits that *everything else also needs to integrate with the reality* (within the design for the unexpected (D4U) applicability range). Admittedly, there exist big dominant systems, leaders in integration efforts, that are not or poorly integrated with reality (e.g., creating their own administrative version of reality). For instance, a tree-shaped bill of materials in enterprise resource planning systems lacks the expressive power needed to capture key characteristics of petrochemical cracking processes, disassembly in the refurbishing of car engines, or cutting operations for sheet metal products. Attempting to deploy them within a context of execution systems is ill advised. Such dominant systems often are old (mostly, they dominate because they were first). As our world changes, these old systems generally are far from optimal today.

D4U goes beyond integration in this leader-follower pattern (with reality as the leader). D4U prefers elements/components/subsystems for which this integration with reality is not only necessary but also sufficient for integration with these elements themselves:

1. D4U preferred system elements are integrated with their corresponding reality.
2. Other components and subsystems are assumed to be integrated with reality.
3. Compliance with 2 suffices to be integrate-able with D4U preferred elements.

In other words, D4U develops components and systems that use (parts/aspects of) reality as a shelter that protects them from (integration) conflicts, even the unexpected ones.

Using *reality as a shelter against unexpected demands* involves a number of challenges. First, D4U elements must not add restrictions, which are absent in the corresponding reality. Nonlinear process plans (Kruth and Detand, 1992) in manufacturing systems constitute a sample technology to answer this requirement. These plans allow to represent multiple manners/options to correctly manufacture an instance of a product type. The precise option that will be selected depends on "unexpected demands" imposed by situations that present themselves (e.g., which machines have free capacity available). Here, every situation may demand a different option.

Answering this first challenge – representing every option that exists in the corresponding reality – often proves to be overly expensive in practice. Hence, a D4U design supports lazy development (=adding options when they are needed). Here, adding options later must not cause a cascade of adaptations to other system elements. The NEU protocol (in the Section "The ARTI Reference Architecture") is a design pattern that precisely addresses this matter.

Second, D4U elements must be able to "keep up with their shelter." Reality is dynamic; it changes in function of time. When D4U elements fail to select a suitable shelter, their shelter will have changed and become ineffective within the time needed to design and develop the D4U element (cf. the parable of the watchmakers in Chapter 4). To this end, a suitable architecture needs to be adopted. The Section, "The ARTI Reference Architecture," presents ARTI, a refinement of PROSA, which is a reference architecture addressing this challenge. The third section, "The DMAS Architectural Pattern," covers the delegate multiagent system (DMAS) architectural pattern, complementing ARTI, further facilitating to cope with this second challenge. DMAS has already been introduced in Chapter 5, which discussed specific DMAS implementations used in holonic execution systems. The section discusses DMAS as a generic and versatile mechanism, identifies which specific DMASs will be present in a holonic execution system in general, and which case/situation-specific DMASs can be developed and used.

Third, the execution system must never fall – virtually – off its world-of-interest. This typically is the least difficult challenge, technically, but requires a proper state of mind. For instance, consider a railway application. There will be D4U software components mirroring railways segments, called "blocks" in railway jargon. Safety measures ensure that more than one train are not present at any time in such a block. However, it is possible – physically – that multiple trains simultaneously occupy parts of a block. If the D4U component cannot mirror its block in a state, when safety measures are malfunctioning or had to be overruled, with multiple trains present at the same time, the execution system will fail to offer support when it is needed the most. It probably will cause the execution system to break down.

Finally, a fully functional execution system cannot avoid including system elements that are not D4U preferred elements. The section discusses such inherently noncompliant system elements (i.e., the intelligent agents), but first the issue of the software/system development process is addressed.

SOFTWARE/SYSTEM DEVELOPMENT

Conforming to the D4U philosophy, this section only presents what is necessary and different in applying D4U when designing and developing software/systems. This can be elaborated further into a comprehensive development process by adopting an existing methodology. The discussion assumes this existing methodology to be object-oriented and incremental (i.e., Unified Process[1] alike).

[1]See en.wikipedia.org/wiki/Unified_Process for more information on this widely known methodology.

Use Cases – User Requirements

In an inception phase, the key requirements for the execution system are elucidated. Typically, use cases are elaborated, allowing all stakeholders to participate and contribute. A use case is a list of steps, typically defining interactions between an actor and a system to achieve a goal. It is a straightforward scenario aimed at simplicity and understand-ability for all involved.

In a D4U approach, these user requirements, use cases, etc. serve to identify and delineate the world-of-interest. They serve to define and specify the problem domain. The requirements themselves (must) remain out-of-sight until later or, actually and ideally, as late as possible. Indeed, user requirements, use cases – except for the identification of the problem domain – must be ignored because (or perhaps more precisely when) they *correspond to the expected*. For example, for navigation applications, a D4U approach must ensure that the solution includes/uses maps, not a booklet with route descriptions for the specific use cases.

Problem Domain Model – The World of Interest

D4U is architecture-centric. From the use cases and key system requirements, an executable model of the problem domain is elaborated. This model will/must adopt the ARTI reference architecture, a refinement of PROSA (see Section "The ARTI Reference Architecture").

By adopting ARTI, the resulting system maximizes the potential for the (critical) user mass of every component or subsystem (see Chapter 4). This reference architecture provides a separation of concerns, as PROSA does, ensuring that the shelters from reality are easy to follow by the components and subsystems. Simple ARTI holons corresponds to parts of the world of interest that do not break into pieces moving in different directions, whereas more sophisticated ARTI holons are aggregates that reconfigure as their real-world counterparts do.

Adoption of ARTI results in a single source of truth (SSOT) design, as PROSA does, in which information is conceptually colocated with its corresponding part of reality. When developing services and other functions on top of this executable domain model, the DMAS architectural pattern is applied to preserve SSOT, even when conceptually global/nonlocal information is needed and used (e.g., when locally selecting a direction at a crossing to reach a remote destination). To this end, the ARTI holons must offer suitable information services mirroring their real-world counterpart (e.g., to answer what-if queries and manage reservations).

Moreover, the domain model explicitly separates pure reality-reflecting parts (of the software) from decision-making elements. The Section, "The DMAS Architectural Pattern," discusses how holons comprise intelligent agents (making decisions) and intelligent beings (mirroring reality). D4U favors to offer functionality through the intelligent beings, maximally restricting the responsibilities of the intelligent agents to the decision-making.

Incremental Development

Modestly recognizing that the human brain is unable to design holonic execution systems in a single design effort, D4U adopts an incremental approach. Every increment results in a working (software) system, which is tested and assessed. From what is achieved and learned, the next incremental development step is planned, initiated, and executed.

An incremental development step comprises the following:

- Delineate and define the domain model to be addressed.
- Design and implement the executable domain models.
- Select the (user) functionalities to be included.
- Design and implement these functionalities.
- Test and evaluate (involving users as much as possible).
- Plan the next step.

As the D4U approach consists of capturing problem domain knowledge, early and significant user involvement is highly desirable. Therefore, elaborating a minimally viable product or MVP (cf. as in the lean start-up[2] approach) is desirable, where viable – for some steps – equals an ability to verify the domain model correctness and adequacy.

At some point, adequacy of the domain model starts to demand completeness: the execution system may never, virtually, fall of its world-of-interest. Typically, this implies that the top-level domain model – corresponding to the world of interest in its entirety – will be an aggregate comprising models for "the remainder of the world-of-interest." These will be coarse models, which require very little effort to develop, providing information services indicating that, normally, human intervention is needed to get the system back to normal operations.

Thus, developers may have to extend their domain model into infinity, in space, in time, in any dimension that is relevant. If the problem domain would be playing chess, the model of the chess board would include

[2]theleanstartup.com.

"outside the chess board," and "add" two rows/columns around the board to prevent a knight falling off the world when virtually moving around. In a manufacturing execution system (MES), the domain model needs to include *a submodel corresponding to the world outside the factory*.

Here, the connections – entrances and exits – between the factory model and this submodel are explicitly modeled. This submodel may be refined at some later point in time, for instance, to account for transportation in networked production where multiple factories operate in a supply chain organization. This submodel may be refined to reflect transportation (times) of products from an exit to an entrance (e.g., when production lines are incapable of transporting products against the normal flow and some products have moved beyond a station that they still need to visit).

Note that (user) requirements (i.e., use cases) are used in later phases of each development step. However, they are to remain out of sight when initiating the subsequent development step. Indeed, especially, the intelligent agents developed to answer user requirements cannot become part of the holonic execution system's core, which is designed for the unexpected.

In Silico[3] Ramp-Up

Because D4U elaborates executable domain models, mirroring the world of interest and offering services to answer what-if queries, the effort needed to have a simulation of a D4U holonic execution system will be small. As execution systems tend to be large (e.g., in steel manufacturing, the MES typically represents an investment that is ten times the amount spent on planning and scheduling systems), a detailed cost–benefit analysis is desirable. And because of its complexity, this detailed analysis needs to be performed on a model/simulation that closely resembles the reality-to-be.

Therefore, a typical approach to handle the transition of the holonic execution system into real-world operations will comprise a simulation phase. The required additional effort will be minimal (cf. Appendix on simulation and modeling). The domain models needed for simulation can be reused without reprogramming for the deployed version of the holonic execution system. At most, some features need disabling. When indicated, simulations with hardware in the loop and humans in the loop will smoothen the transition even further.

[3]Using simulations running in computers (cf. en.wikipedia.org/wiki/In_silico).

THE ARTI REFERENCE ARCHITECTURE[4]

This section discusses the ARTI reference architecture, a holonic reference architecture for *execution systems managing real-world activities using real-world resources*. The reference architecture is both a refinement and a generalization of the PROSA reference architecture for manufacturing systems. To avoid repetition, more detailed information about PROSA may not be repeated here but it remains valid unless stated otherwise. ARTI's design applies and illustrates the principles in Chapter 3 and it accounts for the laws of the artificial discussed in Chapter 4.

Structure of an ARTI System

The reference architecture comprises four kinds of basic component or holons (as it is a holonic system):

- Activity Types (product holons in PROSA)
- Activity Instances (order holons in PROSA)
- Resource Types (parts of a resource holon in PROSA)
- Resource Instances (parts of a resource holon in PROSA)

In addition, there is a class of optional components called *staff holons* (as in PROSA).

An ARTI system is a flexible hierarchy – also called holarchy – aggregating basic and optional components in a possibly time-variant manner. Also, there can be an abstraction relationship among components, where the more concrete member in this relation implements all capabilities of the more abstract component.[5]

Each of the above components or holons, regardless of its kind, is subdivided in reality-mirroring and decision-making parts: an intelligent being (IB) and an intelligent agent (IA), respectively. Intelligent agents may be divided into an instance (IAI) and a type (IAT). Section "The DMAS Architectural Pattern discusses this in more detail.

[4]In software engineering, a reference architecture is defined as a set of coherent engineering and design principles used in a specific domain. It aims at structuring the design of a specific system architecture by defining a unified terminology, the structure of the system, responsibilities of system components, by providing standard templates, components, by giving examples, etc.

[5]The more concrete subclass inherits from the more abstract superclass, when using object-oriented programming terminology. Subclass instances remain superclass instances from birth till death.

ARTI refines PROSA by distinguishing a resource instance and a resource type as two subcomponents of each resource holon. ARTI also makes an explicit distinction between reality-mirroring and choice-making parts within the holons. ARTI generalizes PROSA by adapting a terminology indicating its applicability beyond manufacturing. In fact, PROSA already has been applied effectively, that is, without needing major redesign or modification; in other domains, however, its terminology proved to be ill adapted and confusing in several domains.

ARTI – Basic Components

Activity Types

An *activity type* corresponds to a class of real-world activities. It is knowledgeable about all aspects that are common to activity instances of its type but ignores instance-specific matter (e.g., the state of an instance). A "human" activity type would be considered to be an expert.

In an MES, there will be an activity type for every product model, and this activity type is the information source for process plans, material requirements, etc. Moreover, there will be an activity type for all other activities within the manufacturing system: planned and unplanned maintenance, transport and storage of equipment (e.g., of empty product carriers), setup and changeover of work stations, etc.

In intelligent transport systems, an activity type is knowledgeable about manners to travel, commute, deliver, etc. In a smart grid, activity types are knowledgeable about manners to produce, transport, distribute, store, and consume power, where power is to be understood in a broad sense (e.g., activity types may be knowledgeable about balancing power). Activity types have been investigated, designed, and often implemented for fleet robotics, open air engineering, health care, logistics, etc. (cf. Chapter 7).

The activity type is able to provide information that is *grounded*, which means that (i) this information allows identifying resource types that are able to execute the activity and (ii) these resource types are able to retrieve the information they need from the activity type (e.g., a recipe).

Ordinarily, activity types are nondeterministic and have a lazy[6] implementation. *Nondeterministic* means that an activity type is knowledgeable about

[6]Lazy means that the implementation effort is delivered when the need arises. Alternative manners to execute an activity are made available when a situation presents itself in which this is useful (i.e., the effort required to offer an alternative is likely to be recovered when the alternative is used and, e.g., enables a more optimal resource utilisation, faster activity execution, etc.).

multiple manners in which its activity instances may execute. For example, a commuting activity type may support multiple transport modes – road, rail, etc. – and multiple timing options within each mode.

Ideally, activity types leave all resource allocation options open to their activity instances; nondeterminism is essential to achieve this. In reality, an activity type masters suitable representations of nondeterministic information (to be able to leave all options open for their instances). However, they employ a lazy implementation strategy whenever it is (too) expensive to offer an option without a perspective that they will be utilized. For instance, when offering an alternative to execute a manufacturing activity requires an expensive validation (e.g., produce test pieces) this will only occur when there is an economic benefit that warrants such investment. And, offering another alternative will not force the activity type to forget the current one(s).

Activity Instances

An *activity instance* corresponds to the execution of a real-world activity; a "human" activity instance would be considered to be a manager. It handles all instance-specific information processing and delegates type-related matter to its activity type. An activity instance is responsible for the proper execution of the corresponding real-world activity. To this end, it ensures the necessary allocations of (time slots on) resource instances. It also is the repository for all state information concerning its execution.

However, identifying whether a resource instance is suitable for the execution of the activity is delegated to the types (i.e., activity and resource type determine this among themselves). Likewise, transforming the state representation – reflecting the progress of its real-world activity execution – is also delegated to the activity type (see further: the NEU protocol). The activity instance is also the repository for traces (i.e., data recordings of an activity's history) and it may be archived, when the activity ends, for future use (e.g., analysis when a defect in a product instance is encountered).

In manufacturing systems, there will be an activity instance for every product instantiation activity but equally for maintenance or auxiliary activities. In intelligent transport systems, every commute will have its activity instance. In home automation and smart grids, climate control will have an activity instance.

An activity type – mirroring what is known about its type – is relatively decision-free (except perhaps for their lazy implementation deciding which alternatives to support at a given time). In contrast, activity instances interact, negotiate and decide about their usage of available resource instances. Therefore, their decision-making subcomponent(s) will have significant responsibilities. However, these subcomponents have to remain separate (i.e., modular implementation) and *final* (i.e., reality-reflecting subcomponents must not rely on the decision-making subcomponents for proper operation). More details and advanced features in this respect are addressed below (see section on DMAS).

Resource Types

A *resource type* corresponds to a class of real-world resources. It is knowledgeable about all aspects that are common to resource instances of its type but ignores instance-specific matter (e.g., the state of an instance). A "human" resource type would be considered to be an expert.

In a first implementation, resource types know the technical specification of the resource (e.g., the dimensions of a parking space) and its capabilities (e.g., a railway being able to transport people possibly carrying small luggage). More advanced capabilities of a resource type are addressed later (see Section on DMAS).

In manufacturing systems, there will be a resource type for every type of machinery, both for processing equipment and auxiliary equipment (transport, storage, energy supply, material supply). Moreover, space, human operators and workers, etc. will have their resource type. In fact, valuable capabilities that are in limited supply is what defines what will be a resource in the world of interest. In intelligent transport systems, road segments, crossings, vehicles, truck drivers, trains, parking spaces, etc. will have a resource type. Resource types have been investigated, designed, and often implemented for fleet robotics, open air engineering, health care, logistics, etc. (cf. Chapter 7).

Resource types interact with activity types to discover whether and how a type of resource can be employed to execute one of their activity instances. For instance, an activity type limits a vehicle selection to suitable ones (e.g., big enough) and ensures that only valid routes are indicated as possible transportation routes (e.g., avoid low bridges with a truck).

The information processing capabilities of the resource type determine what *grounded* means for the activity types that consider using it. For example, a car driver may need detailed instructions whereas a railway system or a taxi are able to execute larger-grained tasks. In single systems, this may range from *very concrete and ad hoc* toward *very autonomous and widely applicable.*

Resource Instances

A *resource instance* corresponds to a real-world instance of a resource, which is valuable and has a finite capacity; a "human" resource instance would be considered a manager. It handles all instance-specific information processing and delegates type-related matter to its resource type. It is responsible for the allocation of its real-world instance to activity instances that have the resource instance perform real-world activities. Processing of technical information is delegated to the resource type (e.g., checking whether a product part will fit in its workspace).

The resource instance is the repository for state information. This includes topological information: what are the exits and entries of the resource instance and to which entries and exits – of neighboring resource instances – are they connected? Or is an entry or exit unconnected? Changes in this state are tracked (i.e., in case of reconfiguration). If the resource is composite, what are its components? Moreover, the resource instance tracks status of the equipment itself (on, off, calibrated, etc.) and knows its *visitors* (a product part on a product carrier in a machine; a driver, passenger, or suitcase in a car; personnel in an office; etc.). Again, transformation of state-representing information will be delegated to the resource type whenever this is type-specific.

Similarly to the situation with activities, resource instances are involved in the decision making regarding their allocation. And their decision-making subcomponents must be separate and *final* (i.e., the reality-reflecting part must not need adaptation when it changes). For instance, replacing a first-come, first-served decision-making subcomponent by a priority-based one must be confined to easily identified modules and must not require a modification of a reality-reflecting subcomponent.

ARTI Interactions Among Basic Components

AT–AI Interactions: The NEU Protocol

The NEU protocol is used for the interaction between an activity instance and its activity type. The acronym stands for *N*ext, *E*xecute, *U*pdate. The discussion uses Erlang[7] pseudo-code to denote the software processes' behavior. To be able to understand the pseudo-code, note that:

- Identifiers starting with a capital are variables.
- Identifiers starting with a small character are constants (often called an *enum*); they mainly are used as tags in the messages exchanged by the software processes.
- The "!" operator sends the message after this exclamation sign to a software process identified before this sign.

[7]See www.learnyousomeerlang.com and www.erlang.org for more information.

NEU Phase 1: Initialization

In an ARTI implementation, an activity instance Act_In is created mirroring a corresponding event in the world of interest (e.g., an order arrival in a factory). Act_In receives a reference Act_Typ_ADR to its type Act_Typ at this creation.

The first action of Act_In is to acquire a representation of its state from Act_Typ. To this end, Act_In executes:

```
Act_Typ_ADR ! {init_order_state, Act_In_ADR}
```

Note that init_order_state simply is a tag indicating what is requested. Act_Typ now generates Act_In_State, which contains all information concerning progress made by Act_In. At this initial stage, Act_In_State reflects that nothing has been done so far.

Act_Typ sends Act_In_State to Act_In using its reference Act_In_ADR:

```
Act_In_ADR ! {order_state, Act_In_State}
```

Note that Act_Typ remains agnostic about the instances of its type. To this end, Act_In passes Act_In_ADR to Act_Typ with every request/message.

Conversely, Act_In remains agnostic about the structure and content of Act_In_State. Act_In is the repository for Act_In_State but it systematically delegates any information processing involving Act_In_State to Act_Typ.

NEU Phase 2: Execution

Act_In is coordinating its real-world activity. However, Act_In lacks domain-specific knowledge and is unable to determine what to do. Hence, Act_In instructs Act_Typ to compute which actions are the candidate next steps:

```
Act_Typ_ADR ! {nextSteps, Act_In_State, Act_In_ADR}
```

Act_Typ analyses Act_In_State and compiles a list of valid actions for the next step. This list comprises state-changing actions (e.g., drill a hole or perform a test) as well as auxiliary actions (i.e., transportation or storage) from the perspective of the real-world activity.

Note how Act_In is a manager – in the narrowest sense of this word. Indeed, Act_In remains agnostic about the application domain specifics. Likewise, Act_Typ is an expert, fully ignorant of what is happening in its world of interest. All state information originates from Act_In.

Act_Typ cannot even distinguish whether the state information, provided by Act_In, corresponds to an actual or a fictive state. As a consequence, Act_Typ compiles lists of valid actions that are computed for actual states as well as for the "might be" states in an exploration DMAS or an intention-propagating DMAS (see section on DMAS).

Act_Typ delivers this list of possible valid next actions to Act_In:

```
Act_In_ADR ! {nextActionsList, [Act1, Act2, … ]}
```

Act_In searches, finds and selects a resource instance Res_In that is capable of executing an action Act_x from the list provided by Act_Typ. Concerning an optimized selection of Res_In and Act_x, the reader is referred to the section "on DMAS and on Staff Holons."

Note that Act_Typ and Res_Typ (i.e., the resource type of Res_In) cooperate to determine whether Res_In is able to perform Act_x. Note that Res_In is agnostic concerning the application domain. Likewise, Res_Typ ignores the existence and states of its resource instances. Like Act_Typ, Res_Typ provides its services regardless whether state information about a resource instance is real or fictional.

When Res_In finishes executing Act_x, it informs Act_In about the outcome:

```
Act_In _ADR ! {done, Actx, Res}
```

In order to remain agnostic about the application domain, Act_In request an update of Act_In_State from Act_Typ:

```
Act_Typ _ADR ! {update, Act_In _Stateold, Actx, Res, Act_In _ADR}
```

Act_Typ computes the new state for Act_In and sends it to Act_In:

```
Act_In _ADR ! {order_state, Act_In _Statenew}
```

From here on, the interaction pattern repeats the phase two interactions.

Note that `Act_In` does not select/execute a second action from the list of candidates for the next step. As `Act_In _State` has changed, `Act_In` ignores whether this old list is still valid.

The interaction protocol derives its acronym NEU from this phase:

- `Next    (candidate operations list),`
- `Execute (operation with result),`
- `Update  (order state).`

Phase 3: Finalization

When the list of valid actions for the next step contains `finalise`, `Act_In` may (but must not) select it and perform any wrapping up duties (e.g., archive the trace data). While the list still contains other actions, `Act_In` enjoys a solution space offering alternatives, which `Act_In` may use to optimize its activity. Note that both `Act_Typ` and `Res_Typ` remain unexposed to such activity optimization issues.

Discussion

The above-presented protocol is not mandatory. Some application domain requires a different solution (e.g., in smart grids where time-continuous profiles characterize activities). Some application domains require an extension or enhancement. Nonetheless, the NEU protocol possesses some key qualities that need to be preserved.

First, the NEU protocol creates independence – enabling a separation of concerns – between type holons and instance holons (i.e., between experts and managers). The activity and resource types have dependencies as they exist in reality, which is OK. The activity and resource instances have dependencies as they exist in reality, which is OK. However, the dependencies among types and instances are limited to *generic and need-to-know*.

For instance, the activity type – when applying the NEU protocol – presents a list of possible next steps to an activity instance. The members of this list are not interpreted in any manner by the instance. The activity instance presents this "step information" to resource instances for two purposes.

First of all, the activity instance needs to discover which resource instances are able to execute this step (capability). Second, the activity instance (or its exploring and intention ant agents) needs to discover whether and when such a capable resource instance has the capacity to execute this step, how long it will take and what the expected outcome will be.

In turn, the resource instance will not interpret this "step information" by itself. Instead, it will pass this information to its resource type holon, which generates the answers. These answers are passed to the activity instance holon. At no point, the instance holons require or acquire knowledge about the type-specific aspects.

Conversely, type holons never are the depository of instance-specific information. The state representations, which are computed by these types, reside with the instance holon. These instance holons provide this state information when type-specific processing is indicated. This effectively creates an independence that is relevant in the problem domain.

Second, the NEU protocol prevents exposure to the specifics of knowledge representation inside the holons to the extent that these holon-internal representation may change even at run-time. For example, note that `Act_Typ` may utilize – internally – a nondeterministic representation of all the possible manners to execute activity instances of its type, whereas `Act_In` may use a different (typically simpler) representation. Moreover, both `Act_Typ` and `Act_In` may change their internal representation without the other noticing (except for the content/size of the list with candidate next steps). For instance, `Act_Typ` can be upgraded to support more options, using a more expressive internal representation, without any other component needing to change. Both `Act_In` and other activity types will not notice this upgrade in a manner that affects their software code/correctness. With the Erlang technology, such an upgrade will be possible while the holonic execution system keeps running (i.e., hot software updating).

RT–RI Interactions

Resource instances interact with their resource type to delegate all information processing that is common to the corresponding class of resources. Determining whether an action (e.g., a processing step in manufacturing or passing under a bridge in transportation) is feasible will be delegated by the resource instance to its type. On the other hand, a resource type will require its instances to be the repository for all instance-related information. So, similar to the NEU protocol above, a resource instance will provide any resource state information to its resource type that is needed to compute how an action is likely to perform. Likewise, computing a new resource state representation is done by the resource type employing state information from the resource instance.

Consider a heat treatment furnace. Its resource type (holon) knows how long it takes to raise the temperature from 20°C to 800°C. It is the resource instance (holon) that provides the information that the initial temperature is 20°C to its type. Moreover, the resource instance knows the activity instance, which is the repository for the product state information. When passed to the activity type, the activity and resource type are able to determine that it will be 2 tons of steel that need processing at 800°C during 4 h. The resource type informs the resource instance that it is possible to execute the activity and provides an estimate of how long it will take. When such a heat treatment step is executed, the types update the instance state information, whereas the instances are the repositories for this state information.

The ARTI design allows to use resource types in both a reality tracking mode and exploration or prediction mode. In the first case, the information provided by the instance for the computation of an updated state reflects what really happened (or, more precisely, what the resource instance believes that happened based on its input from sensors and humans). In the latter case, the data for the update is selected, typically by the activity instance (see Section on DMAS), without it happening for real. The resource type has been used to generate the type-related elements of these data (e.g., estimate duration). In fact, the resource instance may have an agenda (see Section on DMAS) and employ its resource type to generate a state trajectory for the resource that corresponds to this agenda.

Consider the heat furnace again. The resource instance (holon) has an agenda containing the processing steps, booked by activity instances, to be executed. From this agenda, virtual execution of the activity instances employ the resource and activity types to compute the estimated future states of both the resource (furnace) and the activities (product parts that are treated in the furnace). The information processing services offered by types and instance suffice to implement this (cf. PROSA discussion in Chapter 5).

AT–RT Interactions

Activity types and resource types interact when they need to provide information, usually to one of their instances, that depends both on the resource type and the activity type. For instance, when the processing time of a production step depends on the CNC program and the machine tool speed, both types will interact to compute this information.

Note that ARTI does not impose which type (resource or activity) must perform specific parts of such computations. In fact, the right choice – in the economic sense – depends on critical user mass and case-specific properties. For example, when a manufacturing process is very common, it will make sense to have the process implementation (i.e., the resource type) offer a lot of services (i.e., have its many users share development efforts). In contrast, when it is an exotic activity, it will be more economical to have the activity type (i.e., a low-level process plan) handle this in an ad hoc fashion (i.e., minimize development efforts).

Resource and activity type interactions will often be triggered by instances and may even occur with instances as intermediaries. Among others, an activity instance will discover what actions may be executed through the NEU protocol and it will contact resource instances to find out whether they are capable of executing one of these possible actions; instances will delegate answering to their type. In contrast, activity types will have a selection of resource types that they will/must consider while elaborating what the possible courses-of-action are for their activity instances. This elaboration may be lazy.

AI–RI Interactions

Activity instances interact with resource instances concerning the execution of both ongoing and planned activity steps. In case of a nonproactive design, the instances have a straightforward behavior: initiate a suitable next action, manage the execution of this action, update the state information, and repeat until the job is done. This is the heterarchical mode of early PROSA implementations.

In a proactive design, activity instances will virtually execute possible courses of action to evaluate alternatives from which they will select one, which is called the intention of the activity instance. Through the virtual execution of this intention, the activity instance informs the resource instances of future usage, which these resource instances reflect in their agenda. This aspect will be discussed in the section on the DMAS architectural pattern.

ARTI: Aggregation, Abstraction, Staff Components

The ARTI reference architecture allows for the construction of large(r) systems through aggregation, identical to PROSA. These aggregates may vary over time when subsystems and components enter, leave, or are replaced. Importantly, these changes and reconfiguration allows to maintain a proper correspondence with the world of interest that is reflected.

Specialization is supported to offer an abstract view on collections of similar but nonetheless different components and systems. Its main purpose and contribution is to simplify the environment of a given component or subsystem when it is not forced to account for nonrelevant differences from its own perspective. Specialization also enables reuse (of software code) but this is considered a minor contribution. Note that using Erlang, specialization is nice to have as it uses duck typing.[8] With static type checking languages, it will be a must-have.

For the introduction of non-D4U systems and components that are not decentralized in a manner compliant with ARTI, the staff holon from PROSA has been retained: staff holons may only have an advisory role. This allows to introduce noncompliant information providers without risking that the execution system breaks down when facing the unexpected.

Examples: Holonic Task Execution Control of Multimobile-Robot Systems

Applying PROSA and DMAS to mobile robots revealed a necessity to translate the PROSA concepts to the robotics world. First and foremost, PROSA terminology is barely appropriate for a robotics environment. The ARTI terminology proved to be adequate and was adopted. Note that ARTI terminology was devised to be broadly applicable, answering the needs of various nonmanufacturing domains while remaining suitable for the MESs.

A distinctive feature of the ARTI/PROSA/DMAS architecture applied to robotics, when compared to existing robot control architectures, is the need for the explicit allocation of resources. The explicit emphasis on reflection of reality, as an SSOT, is another added value with respect to the existing robot control systems. The robotics research community is still struggling to grasp and appreciate the novelty of these ideas, which have been welcomed in the holonic systems research community but have remained unwelcome in the somehow excessively focused/specialized robotics community.

Three cases are briefly treated hereafter. The reader is referred to Philips (2012) and (Huang, 2011) for further details and more cases. The robotic resources used in the use cases are depicted in Figure 6.1.

[8]The name duck typing is derived from the famous quote: "When I see a bird that walks like a duck and swims like a duck and quacks like a duck, I call that bird a duck." In duck typing, a programmer is only concerned with ensuring that objects behave as demanded of them in a given context, rather than ensuring that they are of a specific type.

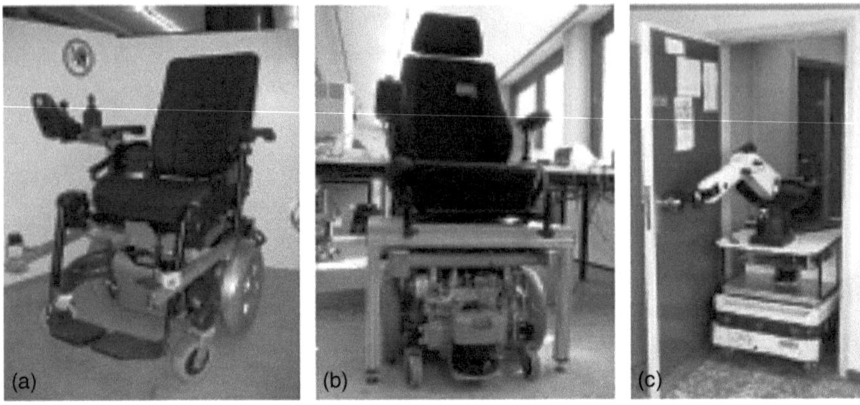

Figure 6.1 ***Mobile Robots Used for the Use Cases.*** (a) SARA, (b) LAURA, (c) LiAS.

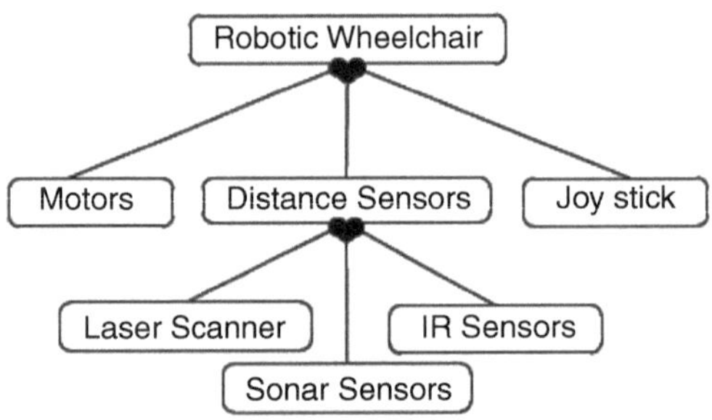

Figure 6.2 ***Resource Type Holons for a Robotic Wheelchair.***

Case 1: Single Robot Allocation in General

At the level of a single robot, the main robot components constitute resource holons. Figure 6.2 shows resource holons inside a mobile robot such as the robotic wheelchairs depicted in Figure 6.1. The level of detail is task dependent, because only resources with an active role need to be represented. If, for example, CPU optimization is required, one of the resources would be the CPU. By doing so, order holons can explicitly allocate CPU time slots, for example, to guarantee real-time performance.

The Robotic Wheelchair resource type has knowledge about its dimensions, its overall minimal and maximal speed and capabilities such as obstacle avoidance and navigating from one point to another. It consists of motors controlling translational and rotational velocity, Distance sensors scanning for obstacles and a joystick representing the user's input. Typically, several distance sensors are mounted on a mobile robot. In this case, a laser scanner, sonar sensors, and IR sensors are represented by resource holons. These sensors are used to detect obstacles in the environment and provide the input to, for example, an obstacle avoidance algorithm.

At the single-robot level, activity type holons and activity instance holons can be defined that address primitive actions such as moving the robot in a given direction while keeping a safe distance (as measured by the available sensors) from obstacles. ARTI implementations enable to realize such a task as the composition of two more primitive tasks: move to a target position and avoid obstacles. Note that this implies a resource allocation in which two tasks own the same resource at the same time but with different ownership rights. These rights are compatible and, normally, complementary. For instance, the *move to target position* activity has received the right to steer the mobile robot but the *avoid obstacles* activity has the right to overrule, modify, and constrain what the mobile robot resource instance may execute.

Moreover, an ARTI implementation makes it possible to allocate these embedded resources to tasks/activities from outside the single robot. For example, one robot's laser sensor may be used by another robot or another type of activity. This other activity could be a building mapping task. Note that this mapping can have different requirements and outputs depending on where it resides. Every building type (factory, hospital, corn field, copper mine) and every mobile robot manufacturer may have its own map-building activity holon type.

These mapping activity types undoubtedly have companion resource type holons for the processing of sensor data. In the example, the laser sensor will have a basic/primitive resource type that will be used by these more sophisticated companion resource holons. In these situations, the split of the PROSA resource holon in two ARTI resource holons (type and instance) ensures useful maneuvering space to combine holons into suitable time-varying aggregates. In case of PROSA, the use of specialization – combining type and instance functionalities in a single resource holon – remains possible, which may imply annoying constraints (when life cycles including creation and hibernation/destruction have to coincide).

D4U compliance involves explicit resource instance allocation (second design principle) to have, among others, an upper bound on the inertia of design choices that need undoing. It also implies minimizing the needs/requirements for resource allocation from activities offering a service (first design principle). A D4U-compliant activity:

- does not request allocations that it does not use. In an industrial robot, the activity (program) that is executing will have all resource instances of the robot allocated (as there is no explicit allocation management).
- does not request allocations for more time periods than needed. For instance, access to sensors via a sensor bus often supports isochronous channels, each providing hard real-time bandwidth to the sensor readouts. A D4U-compliant activity will acquire such channels (= resource instances) that it needs, leaving the remainder for other activity instances.
- does not request more rights than needed. For example, a safety-ensuring activity needs isochronous channels to the sensors it needs to see imminent danger, hard real-time CPU slots to process the sensor data and assess the need for intervention, and preemption rights on actuators/power/brakes to intervene when indicated. This leaves to other activities sufficient rights to execute the actual robot tasks. It enables a separate safety-ensuring activity to be provided regardless of the nature, source, etc. of these other activities executing the application tasks.
- does specify which alternative resource allocations exist/are possible. Moreover, such nondeterministic specification may describe the trade-offs between allocation and task performance, capabilities, etc.
- does minimize the requirements for allocation and deallocation. For instance, a special-purpose trajectory execution activity maximizes the robot state (e.g., positions and velocities) in which it can perform a handover from/to other trajectory execution activity types.

Note how explicit resource allocation captures the interactions among tasks/activities. This SSOT solution helps to cope with the complexity of combining a multitude of tasks in a robot system while maximally utilizing the available resources.

Case 2: Obstacle Avoidance

In mobile robotics, the goal of obstacle avoidance is generally to navigate from one location to another while avoiding collisions with obstacles in the environment. These obstacles can be structural or static, such as walls or doors in the environment, but also dynamic such as other robots or humans moving in the same environment.

Often this problem is solved in the control algorithm of the robot in a reactive manner, in contrast to (global) path planning where a precomputed obstacle-free path is followed by the robot. The robot controller reactively

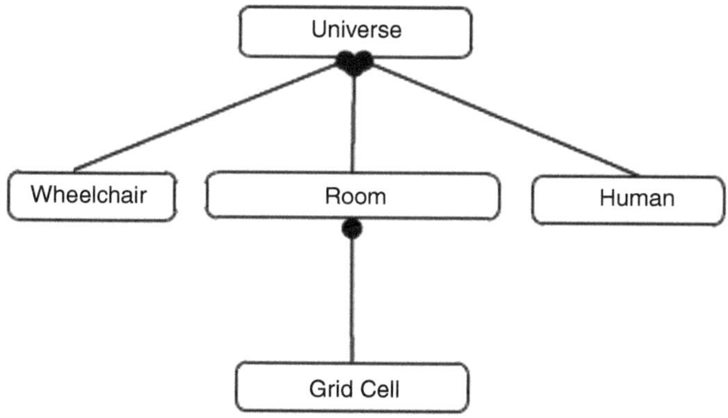

Figure 6.3 ***Resource Type Holons in Obstacle Avoidance Task.***

alters the robot's course to avoid the obstacle and afterwards steers back to the original trajectory.

The obstacle avoidance problem can be interpreted as a resource allocation problem in which the environment is divided into a set of grid cells and the desired trajectory is represented by the allocation of a sequence of these grid cells. The resource holons representing the environment are, thus, divided into grid cell resource holons that can be individually allocated. Figure 6.3 shows the relevant resources in this use case.

The mobile robot is allowed to move along its desired path if it allocates the required sequence of grid cells at the relevant time intervals. Another robot moving in the same environment will not be able to allocate the same space at the same time, and this way collisions between both robots are avoided. The checkerboard cells are allocated by the robot during its trajectory.

These allocations are temporal, and the time interval of each allocation should represent the time when the robot actually occupies this location. This solution, however, assumes that all actors in the environment adopt this resource allocation. It does not take into account humans or other robots moving around without allocating their space beforehand. Another problem might be the difference between allocation time and actual execution time of the trajectory. If the model, used to estimate the time a robot needs to move to its target, is not adequate, robots might deviate from this schedule.

The short-term forecasts provided by DMAS are able to detect such deviations and, if sensor data are linked to the environment resources, dynamic obstacles can be taken into account as well. Nevertheless, how fast the robot is able to avoid obstacles depends on how fast a change in the environment is detected. If the robot only relies on its own sensors, forecasting is very limited and there might be not enough time to complete a full exploration and allocation cycle using DMAS. If, however, each room is equipped with its own sensors to keep its occupancy up-to-date, changes that are in conflict with a robot's trajectory can be propagated in time, through the exploring DMAS, and alternative solutions can be selected. Likewise, all sensors of all robots can be used, updating the resource holon corresponding to grid elements. The DMAS will inform all activities affected by information from these sensors.

Moreover, when intention ants propagate the intentions of their mobile robot, proactively, the resource allocation of grid elements over time will be visible and can be managed. Among others, deadlocks will be prevented and travel times/distances can be minimized.

Case 3: Multirobot Door Opening

This type of application requires to include the environment of the robots explicitly and have resource holons representing the resources therein. The goal in the multi-robot door opening task is to have one or more (mobile) robots to navigate from one room to another through a door opening. Some of these robots are able to open and close that door. Ideally, if more than one robot has to move through the door within a short time interval, the door should be opened only once.

Figure 6.4 shows the resource holons for a door-opening scenario/task. The Universe resource allows future additions of resource holons to the application. Here, the Universe is divided into resources belonging to the mobile manipulator executing the task, task-relevant resources such as the Door, its Handle and its Frame, Robotic Wheelchair resources, and the Human operator whose control is shared with the wheelchair.

This explicit resource representation and resource allocation in the robot's environment facilitates coordination. For example, when two wheelchairs arrive at the same time at the door, the explicit allocation of the door resource ensures that the mobile robots are able to exit one by one and to

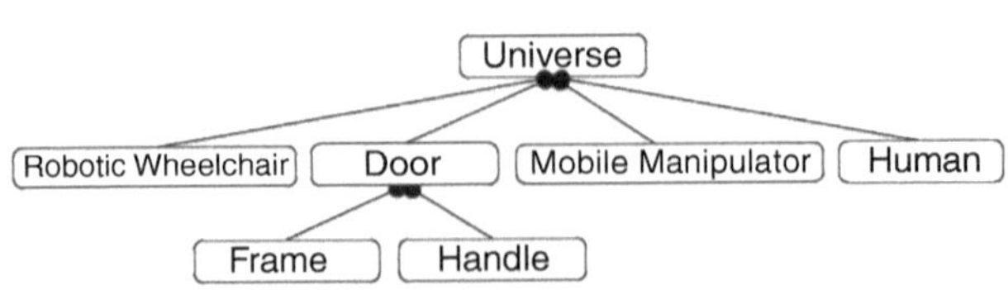

Figure 6.4 ***Resource Type Holons for a Door Opening Task.***

avoid deadlock. Assume that both wheelchairs are driving toward to the door opening and issue allocation requests to the Door Resource. Since these requests are handled sequentially, the first robot requesting allocation at a particular time, will receive that allocation slot, while the other robot will receive a slot behind the previous slot.

If both wheelchairs reach the door within an acceptable time delay, the door should only be opened and closed once. The door holons will handle this in a manner analogous to a machine holon in a factory managing its changeover and setups. When activity instance holons handling the wheelchairs apply the DMAS mechanism, the resources in the environment will be informed about future allocations. Therefore, the door opening coordination is not limited to handling the current situation. The order holons will observe the predicted future states of the resources in the environment and are able to account for these predictions. For example, when an activity instance holon sees that it will have to wait for other wheelchairs traversing the door in the opposite direction, it may adapt its trajectory in space and/or time, in order to reduce power consumption, robot component wear, or have its occupant wait in a more agreeable or useful location (e.g., in front of a television or at a battery recharging station).

THE DMAS ARCHITECTURAL PATTERN

The delegate multiagent system, an architectural pattern, has been introduced in Section "Bioinspired Coordination and Control in Holonic Execution Systems" of Chapter 5. This section revisits the DMAS pattern and describes it as a generic problem-solving mechanism and its roles within the consolidated research results. It presents what a DMAS minimally involves; it presents which DMAS will be contributing what in an ARTI-compliant holonic execution system; it hints at the variety of DMAS implementations that can be present in an execution system; it discusses the intelligent agent's role as a "plug-in" components to isolate the nonD4U elements from the D4U elements. Thus, this section presents a systematic view on DMAS within a holonic execution system.

Recalling Section "Bioinspired Coordination and Control in Holonic Execution Systems," of Chapter 5, a DMAS involves a holon creating, at regular time instances, lightweight agents. These agents are called *ants* to honor the biological source of inspiration, which allowed for the discovery of this pattern. These swarms of digital ants provide services to their holon and/or the overall system.

The Barebones DMAS

A DMAS allows to process "global information" while preserving an ARTI design in which real-world counterparts have an SSOT. The SSOT is provided by means of holons mirroring resource instances, resource types, activity instances, or activity types, including the structural aspects (aggregation). The results of such nonlocal information processing are made available locally, typically linked to a suitable source of truth (stigmergy).

Biological ants deposit chemical/pheromone trails on the physical surface of a real-world environment. These trails inform locally about a nonlocal "truth" (i.e., in which direction there will be food). These trails have complicated geometries without the need to mirror this within the brain of the ants. The world itself is used as its own model. In a similar manner, DMAS uses its ARTI mirror image of the relevant reality. The digital ants travel across this digital mirror image, observe it, influence it, and deposit information on this virtual image.

Furthermore, the evaporate-and-refresh in the biological world has its counterpart in the ARTI image. Information (a digital pheromone trail) has a finite life span and will be redeposited regularly if it is to remain available. This enables to cope with changes in the world of interest.

Generically, digital ants behave as follows:

- An ant is created by a holon, which provides a procedure (code and initializing data) to execute.
  - The data contain the address – used to send messages – of the creating holon.
  - The data contain the address of the current location/holon for the ant.
- Either, the ant processes information on its current location.
  - The procedure provided by the creating holon determines what happens.
  - This can be observing the current holon.
  - This can be influencing/informing the current holon.
  - This can be depositing digital pheromones.
  - This can be observing digital pheromones.
- Or, the ant travels to a neighboring location.
  - It sends a message to the current holon/location to receive a list of connections.
  - It selects a connection and – virtually – travels to the connected neighbor.

The general scheme above has no restrictions except for the ants to be lightweight and self-contained agents:

- Every basic step in the above procedure, an interaction with a holon is bounded in computational effort, time and memory. Ants may never invoke services – from holons or otherwise – that have undetermined requirements on computational resources. For instance, ants may not invoke an optimization service that iterates until its solution converges without an upper bound on this number of iterations.
- Every ant has a *hop limit*, which is the maximum number of basic steps that it may execute. Moving to a neighbor counts as a basic step. Invoking a service from a holon also counts as a basic step.
- The DMAS itself has an upper bound on the frequency at which ants are created.
- Ants must discover information starting from their initializing data. For example, an ant may send data to the creating holon at any time, may recall the addresses of holons that have been visited already, may utilize information deposited by other ants, etc. However, the ant may not rely on global information unless provided by the initializing data.

Research has investigated an optimizing variant: the cloning ant. When the ants of a DMAS would duplicate efforts by repeating the initial part of their virtual journey, it is more efficient to have a single ant perform the shared part and have this ant clone itself whenever the journeys (virtual trajectories) start to differ. Here, an ant will be created with cloning budget, which is an upper bound on the total number of copies that can be made. When cloning, this budget has to be divided across the copies (i.e., the cloning budget is not copied). Overall, DMAS designs and implementations are guaranteed to be (computationally) efficient; thus, they are bounded effort, variable result service providers.

Intention-Propagating DMAS – Predicting the Unexpected

As discussed in Chapter 4 (Law 4), imagination[9] is needed for a system to be proactive in a complex-adaptive world (in which knowledge of past behavior no longer suffices to estimate the effectiveness of an envisaged action). This imagination needs to be collective for many of today's challenges.

The main mechanism to generate predictions (i.e., to create this collective imagination) when knowledge of past behavior is insufficient, is

[9]Not the imagination of a small child fantasizing about something that is impossible but the imagination of an adult picturing what may or will happen and which is used for planning future activities, e.g., to imagine how long it will take to drive to Brussels during rush hour.

the intention-propagating DMAS. Here, activity instances (holons) create intention ants. Intention ants virtually execute their activity instance's intention and, thereby, inform resource instances (holons) of their activity instance's intention to use them in the future.

This intention propagation executes as follows:

- Every activity instance has decided how it will execute (= the intention of the activity instance). The DMAS itself is agnostic concerning how the intention of the activity instance is obtained.
- The activity instance creates intention ants at the location(s) of its current state, equipped with a procedure to virtually execute the remainder of its activity along the current intentions.
- Each intention ant virtually executes the remainder of the activity:
  - Executing the NEU protocol to verify whether the current intentions are still feasible. Recall that activity types cannot distinguish whether they are used for virtual or real-world execution purposes.
  - When infeasible, the intention ant reports the extremely poor estimated performance to the creating holon (activity instance) and dies. As the intentions will not be refreshed, the involved resource capacity reservations, made for an infeasible intention, will "evaporate."
  - When feasible, the intention ant virtually executes the next step of the activity on the resource instance(s) indicated by the intention of its creating holon (activity instance).
  - This virtual execution informs the affected resource instance(s) of the intention and, as a result, makes a reservation for the required resource capacities.
  - This virtual execution invokes services from the affected resource instances to generate an estimate of how the step execution will perform (timing and outcome). The resource instance is likely to invoke services from its resource type and possibly the activity type to compute the estimated outcome.
  - The estimated outcome is used in the NEU protocol to update – virtually – the state representation of the activity instance. Subsequently, the intention ant is ready to execute the next activity step of the intention (virtually).
  - An activity step can be an actually processing step but equally a transportation or storage step.

Note that the intention-propagating DMAS requires that activities and resources are knowledgeable about themselves in a what-if mode. Importantly, this knowledge remains self-knowledge (i.e., local), preserving the

SSOT characteristic. Also, it is self-knowledge that remains valid when the predictions – generated by the intention-propagating DMAS – are made available throughout the holonic execution system. Furthermore, resource instances have an agenda functionality to register reservations. This functionality serves to accurately predict the outcomes of activity steps, accounting for the contention among activity instances for the limited availability of resource capacities. Likewise, activity instances have functionality to represent their intentions such that intention ants may execute them virtually.

Recall how the discussion of the PROSA reference architecture in Chapter 5 did address this already. ARTI preserves this while increasing reusability when resource instances, providing the agenda management functionality, can be combined with numerous types of resources/equipment.

Summary

Resource types are enhanced by self-models to simulate the execution of activity steps. This simulation allows to compute the required properties such as processing time and possible outcomes (with their probability). Note that these self-models often will be very simple (e.g., time to make an omelette plus the probability of success versus failure).

Resource instances will have an agenda containing the foreseen activity steps. The self-modeling capability of their resource type permits to compute the corresponding state trajectory of the resource instance.

Activity types, when executing the NEU protocol, are incapable of distinguishing being used in a simulated or real mode. As they use their instances as a repository for state information and as a source for progress information (what has been done with what outcome), an activity type has no way of knowing whether the provided information corresponds to the real-world counterpart or virtual execution.

Activity instances are able to represent their intentions, how they plan to execute, in manner that allows their intention ants to execute virtually whatever remains to be done.

The Exploration DMAS – Decentralized Search

The above collective imagination, generated by computing how the future will look like if all the prevailing intentions are executed, needs to be used to improve the coordination in the holonic execution system. The prediction allows to observe issues and opportunities in time to do something about or with them. This section presents a generic DMAS to explore for solutions – possible ways to execute activity instances – and their estimated performance/behavior. It is called the exploration DMAS, which is composed of exploring ants and created by activity instances (holons).

The exploring ants behave similar to the intention ants: they virtually execute (the remaining part of) an activity to compute/predict/estimate its behavior and performance. The key differences are:

- exploring ants do not make reservations. They only make inquiries with the resource instances about estimated performance and outcomes without declaring an intention to actually execute the activity steps identified in the interaction.
- exploring ants are equipped with decision-making mechanisms, provided by their creating activity instance, to steer their search. Intention ants receive the intention of their creating holon, fixing by definition all decisions needed for virtual execution. Exploring ants have a search-steering mechanism that determines which options they will investigate.
- exploring ants use this decision-making mechanism to:
  - select a member from the list of candidates offered in the NEU protocol,
  - select a resource instance to execute this selected member, and
  - overall, to select a trajectory among all the possible ones.

Typically, an activity instance utilizes a mixture of exploring ants, mixing decision-making mechanisms. The following DMAS concepts may, for instance, be implemented:

- Candidate solution refresh. When the activity instance is about to change intentions (see below), it first refreshes any candidate solutions, which are considered to become the new intention. When the estimated performance for the current intention, reported by the youngest intention ant, has deteriorated because of a disturbance, it is advisable to verify whether their replacement is not affected as well. This exploring ant behaves precisely as an intention ant except for not making any reservations with the resource instances.
- Random search. The exploring ants simply select randomly a candidate activity step or movement onto a neighbor. In a manner, this is the ultimate in robustness in the sense that it makes no assumptions. Unfortunately, these ants are unlikely to discover good solutions when there are many options and selecting a proper one has a significant impact on performance.
- Randomized search. The ants make biased randomized choices where the bias is a heuristic favoring the more promising selections. By keeping

all options – asymptotically – open, robustness is again preserved. If the heuristic faces unexpected conditions, rendering it contraproductive, options that are noncompliant with the heuristic may still be discovered.

- Shortest path. The exploring ants are assisted by a staff holon that executes a shortest path algorithm. They follow the shortest path, possibility managing/exploring activity steps that are neglected by the staff holon.
- Schedule execution. Similar to the shortest path but the path is computed by a scheduling staff holon. This was discussed in Section "Cooperation of HMES with Planning Systems" of Chapter 5.
- Track record / common practice. The activity instances leave a digital pheromone trail indicating the path that they have followed, indicating their class (e.g., their activity type). Exploring ants are attracted to this trail, similar to the shortest path.
- Bottleneck avoidance. The prediction from the intention-propagating DMAS is used to identify the location and time of the bottlenecks in the system. The ants will prefer to avoid these bottlenecks.
- Aggregate demand profiles. This requires cooperating DMASs. A first DMAS propagates demands where both slack and the availability of alternatives translates in a less than 100% (imaginary) load on resource instances. For example, if 200 minutes are available for 100 minutes' processing, the resources load is 50% during 200 minutes. Similarly, if four resources are available, they are each loaded for 25%. All loads are added, resulting in aggregate load profiles, which may exceed 100%. The second DMAS comprises ants preferring solutions that avoid the peaks in the aggregate profiles.
- Alternative solutions. This DMAS complements the other ones that are present. It requires the other exploring agents to leave a pheromone trail. The alternative-exploring ants prefer solutions that are different. The ants are likely to discover solutions that remain unaffected by disturbances invalidating the more ordinary solutions.
- Continuity of care. In situation where activities are repeating (e.g., in home nursing), solutions employing the same (human) resource for an activity instance often will be preferred. Indeed, this avoids communication problems (when nurses visit a patient treated by a colleague). Continuity-favoring ants will search for solutions employing the same resource rather than hopping around, without getting stuck when continuity cannot be preserved.
- Batching. A DMAS may propagate "resource intentions," making batch properties and especially batch switching visible in a virtual

neighborhood of the resource. Batching ants will aim to join suitable batches or, otherwise, insert a new batch where a switch will happen anyhow.

- Returning. The ants collect information on the first half of their virtual journey, process this information, and finish the job on the return trip of their virtual journey. This can be necessary/useful when the resources exhibit constraints that make it difficult to select a feasible timing, slot, etc. without information about the entire journey.

Clearly, the possibilities for designing an exploration DMAS are practically infinite. Because of the ARTI image and the intention-propagating DMAS, implementing an exploration DMAS requires relatively little effort. Case-specific designs are to be used and generalized when sufficient lessons have been learned.

Importantly, the holonic execution system may/will combine multiple exploration DMASs. Together, they deliver the activity instance a collection of candidate solutions. From this collection, the activity instance selects its intention. And, when time catches up with the intention, its execution is initiated (to be performed by resources).

Intention Selection

The activity instances use a combination of exploration DMASs to maintain a collection of candidate solutions. Membership depends on the performance estimations for the solutions, and possibly, complementarity (contribution to robustness). The intention of the activity instance is selected from this collection, and its intention-propagation DMAS informs the affected resource instances while re-computing through virtual execution the estimated performance and behavior of this intention.

The initial intention selection depends on the proximity, in time, of the first activity step (i.e., the resource needs a minimal amount of advance warning) and the performance improvements of solutions discovered by the exploration DMAS (i.e., if it converges/stagnates, an intention is selected). Changing intentions has been already discussed in Section "Socially Acceptable Behaviors for Delegate MAS" of Chapter 5.

Other DMASs

In Chapter 5, the feasibility DMAS was presented. However, after applying D4U in multiple application domains and a wider range of manufacturing systems, it became apparent that this feasibility DMAS is not universal to the same extent as the intention-propagating and exploring DMAS. Its nature will vary depending on the application domain and even application. What remains is that the exploring DMAS needs to be able to supply an adequate amount of high-quality candidate solutions. A feasibility DMAS ensures that unfeasible options are not explored. However, it is not an issue if a small percentage of exploring ants fail while attempting to virtually execute an unfeasible journey. Feasibility DMAS must not be perfect in some cases. In other cases, it may indicate distances when activity instances may travel virtually from any location to any other location (and all routing choices will be feasible).

Moreover, in the discussion of the exploration DMAS, supporting DMASs have been introduced: (i) bottleneck identification, (ii) resource intention propagation, and (iii) aggregate demand profiles. The delegate MAS architectural pattern is polyvalent. In particular, collecting information about a system (state) property and propagating this across the system will provide services that comply with D4U. Much remains to be explored in this respect.

Intelligent Agents Versus Intelligent Beings

The above reveals that the main decision-making happens in the exploration DMAS and the intention selection. As stated in the Section, "The ARTI Reference Architecture," ARTI holons and other components explicitly separate into intelligent beings (mirroring relevant corresponding reality) and intelligent agents (with the decision-making responsibility). Here, it is possible to distinguish the following intelligent agents:

- Activity instance intention selection. The agent embodies the selection criteria and procedures deciding:
  - when to select the initial intention,
  - which intention to select, and
  - when to switch intentions.
- Membership of the collection of candidate solution. The agent decides whether a solution is retained.
- Resource instance policy. When receiving reservation notification/request from an intention ant, the agent determines how this will be honored. For instance, a request compliant with scheduling advice may push another reservation to a later point in time.
- Exploring ant decisions. The mechanisms determining which part of the solution space will be explored are implemented in an intelligent agent.

These intelligent agents are "plug-ins" in an ARTI software platform, which are noncompliant with D4U. When they are replaced, updated, etc., this must not cause a cascade of required changes.

Moreover, because the DMAS pattern uses virtual execution, an intelligent agent can be seen as a digital mirror image of itself. It is part of the executable domain model. When a decision mechanism changes, its model changes and the DMAS virtual execution adapts to the changed situation with requiring any update or software maintenance for itself. However, when addressing challenging real-world cases, not all intelligent agents can be their own model. When they are computationally expensive or decisions are made by humans, separate reality-mirroring models will need developing (cf. Section, "Toward a Humane Mechatronic Society").

CHALLENGES AND LESSONS LEARNED FROM APPLICATIONS

The above sections discuss consolidated research results. However, when applying these results to real-world manufacturing[10] or to new application domains, new challenges and issues arise. They do not break or invalidate the consolidated results but present requirements for additional services and functionality, or they open discussions on terminology and their understanding by the concerned communities. This is addressed within this section.

Interoperability

Design for the unexpected, aiming to address integration issues, sheds a new light on the concept of interoperability. In the perspective of a profound understanding of integration-ability, the applicability range of interoperability becomes more sharply delineated.

In the research community investigating interoperability, interoperability connects existing systems as they are. These connected systems collaborate without changing themselves in an in-depth manner. In this research community, system integration involves in-depth adaptation resulting in a new, larger system. The systems that are integrated are more or less redeveloped and often are no longer capable of surviving outside the larger system in which they are integrated. This picture may be exaggerated to enhance the contrast with a D4U view on this matter.

To integrate D4U designs, it suffices to render them interoperable. To obtain a larger integrated D4U system, a majority of D4U elements is made

[10]Invitations from and/or preparedness of manufacturing companies for joint research invariably involved cases that could not be handled by existing commercial solutions; some unexpected challenges had to be tackled.

interoperable while the adding/developing a minority of non-D4U elements completes the development of this larger system.

The above should trigger a discussion on what interoperability and its accompanying standardizations may contribute. In-depth understanding reveals that:

- When interoperability and standards remain decision-free (i.e., D4U-compliant), they are likely to be successful and easy to implement. For instance, converting Cartesian (x, y)-coordinates into polar (r, Θ)-coordinates is D4U-compliant. In contrast, scalar noise measurements, that is, weighted sums assigning importance to the contribution of the respective frequencies (like dBA measurements), involve decisions. They may not so easily be exchanged among systems. For structural noise isolation in buildings, low-frequency contributions are likely to be underestimated. The music industry may maximize damage to a person's ears within the legal constraint imposed by such a measurement. Here, there is no guarantee that interoperability efforts will succeed in bringing good value to the users.
- Large margins may render such decisions, needed to render systems interoperable, easy to ignore, or almost irrelevant. For example, when noise limitations can be very strict, the weighing becomes a nonissue. Likewise, representing time in milliseconds or even microseconds by 64-bit integer values is a good idea (from a D4U perspective), even when the applications consider 10 min to be its smallest relevant chunk of time when coordinating their activities. Indeed, consider three systems managing time in 15-, 20-, and 50-min timeslots (e.g., for teaching). The smallest common multiple, when they are to collaborate, is 300 min. Here, the second D4U principle would have limited the inertia of the decisions to use such large timeslots. Typically, the decision-making subsystems (the intelligent agents) dislike large margins when they inherently cause their solution space to explode. The reality-reflecting subsystems (intelligent beings, visualization, and user interfaces) have no problems with these large margins. D4U design principles will minimize the amount of work that would be needed to solve the issues with the larger timeslots (small margins). Only some of the intelligent beings would need to be adapted.
- D4U avoids being penny-wise and pound-foolish. The above choice for a large margin corresponds to what is easy to implement on today's and tomorrow's computer systems. It is easy to generate time measurements in microseconds; it is easy to synchronize clocks on a millisecond level; it is easy to use 64 bits to represent a time instance; and 64-bit suffices to

cover the foreseeable future (i.e., more than 5000 centuries when using microseconds, five million centuries when using milliseconds).

- D4U design decisions first look at what is possible without a cost that might be considered prohibitive (i.e., they are pound-wise) and use that to maximize margins (i.e., they are not influenced by penny-sized gains). The designers look at what is affordable first and to what is really required second. They select what is affordable even when it is overkill for what is needed.
- A lot of systems are able to interoperate because they are able to squander bandwidth and/or memory space when they are using XML, JSON, etc. to exchange information. This is a widespread illustration of how large margins render interoperability to be feasible.
- Many interoperability and standardization efforts are doomed to fail. Here, D4U insights will reveal that there are inherent conflicts and how serious they are. A common symptom of these situations is a struggle for control.
- In addition, there may be the perception that yielding control to another party, and complying with the standards and application programming interfaces (APIs) proposed by this other party, will not solve the issues. Yielding control simply results in a never-ending effort to comply with an ever-changing standard and API definition under the control of this other party. In other words, the parties involved do not believe there can be a stable interoperability standard.
- In fact, these parties are calling and labeling something to be a standard that, by definition, is not a standard at all; they simply are fighting for control and abusing the respect for standardization efforts (e.g., by government representatives controlling their funding).

Overall, this discussion still needs to be continued. It is too early to present conclusions. However, it should be apparent already that way too much value and in-depth contribution to society is expected from interoperability and standardization efforts. Their contributions and value probably may be limited to addressing *last-mile gaps* for which straightforward developments and standardizations suffice while all in-depth issues need D4U. Furthermore, standardization and interoperability also concerns business and legal considerations (e.g., to prevent proprietary solutions from becoming gatekeepers) outside and beyond the present discussion. In other words, this is to be continued.

Multiresource Allocation and Auxiliary Operations

Many real-world problems involve multi-resource allocation. In a holonic logistics execution system (HLES), transportation activities require a truck, a truck driver, a docking slot, a forklift, a forklift driver, etc. In open air engineering applications, a rendezvous has to be arranged between a combine harvester and a tractor, between an asphalt layer and it supply truck, possibly managing drivers for them as well. Similarly, auxiliary operations need coordinating with the main/actual operations. Trucks need refueling, drivers need to rest, machines need cleaning, changeovers, etc.

Modeling these situations is adequately addressed by the consolidated research results discussed above. However, facilitating and supporting the coordination and decision-making has been addressed on a case-specific basis within the research activities until now. It remains an open question to what extent such ad hoc problem solving, current practice in operations research, is inherently unavoidable. Nonetheless, a collection of solution templates can be developed.

Three classes of solutions have been identified within the case-specific developments. First, there is the leader-follower class. Here, resource allocation of the leading resources happens along the approach discussed above. From this allocation, the remaining resources are allocated. This assumes that only the leading resources are scarce while the follower resources are plentiful and almost always available.

A second class allocates these nonleading resources through fixed schedules (e.g., personnel operating the machines, planned maintenance), by providing every leading resource with its nonshared nonleading resource (e.g., a measuring probe), by assuring a supply of consumables, etc. This is the most common situation in practice. It can be made more adaptive and responsive when a holonic execution system has been deployed.

The predictions generated by the intention-propagating DMAS can be used to schedule and plan these auxiliary operations (e.g., preventive maintenance when a machine would be idle anyhow, make sure components are fed where they are most urgently needed). When, for example, a measuring probe on machine A gets damaged, the predictions allow to assess whether it is a good idea to remove an identical probe from machine B and put it on machine A.

Conversely, when these auxiliary operations fail to keep a leading resource fully operational, this will be detected early (by the intention-propagation DMAS of the auxiliary operations). Intention ants and exploring ants will observe this unavailability (temporary and possibly partial) of the leading resource and adapt. Here, holonic execution systems provide visibility, including prediction, which facilitates multi-resource allocation and handling of auxiliary operations.

A third class addresses nonleading resources that are very similar and available in some numbers. Transporting devices are a typical example. These resources can be placed in a pool. A pool holon manages their initial allocation as well as reallocations. When an activity needs a resource from a pool, it contacts the pool holon to get a specific resource assigned. Detailed interactions occur between the actual activity and resource holons. This pattern still needs further investigating. Can the pool holon be a staff holon? Is it possible and advisable to change an assignment (e.g., when a truck breaks down)?

Overall, the predictions offer interesting possibilities while further investigations, involving real-world challenges, are needed to consolidate results concerning multi-resource allocation and handling auxiliary operations.

Probability (Distributions)

In real-world operations, the expected outcome of an activity step will be stochastic. In electronics, the expected outcome of a (binning) test might be:

- 7% of the integrated circuits functions above 2.9 GHz (clock rate),
- 43% of the integrated circuits functions above 2.5 GHz (clock rate) but fails at 2.9,
- 33% of the integrated circuits functions above 2.1 GHz (clock rate) but fails at 2.5,
- 14% of the integrated circuits needs disabling of a CPU core and retesting, and
- 3% of the integrated circuits needs to be scrapped.

The production of optical lenses or of bearing balls may have a similar characteristic.

Contributing to a sustainable society, disassembly and recycling, repairing or refurbishing become increasingly important and prevalent. These activities also have the above characteristic when, after disassembly, tests reveal whether a component can be reused as-is, needs some further processing, or has become scrap material.

To cope with the above, the intention propagation DMAS needs to master probability distributions. Instead of propagating and transforming scalars (e.g., expected arrival times), it uses a suitable probability representation. Note that this is likely to use cloning ants where each clone continues along a possible future routing of the activity (e.g., one ant for the high-value packaging of an integrated circuit, another for midrange, another for trimming a CPU core, etc.).

Propagation stops when hop limits are reached and/or cloning budgets get exhausted, or when the probability information no longer contains usable information (e.g., when aggregated predictions based on historical data – predicting the expected – deliver equivalent value/information).

No research activities have addressed this until now. Note that this does not require to adapt the consolidated results at all. It suffices to enrich the information handled by the ant and holon without any "structural" changes. Conceptually, concerning D4U, this is not a challenging task.

Also note that both the scalar and the probabilistic DMAS can coexist; the designers of a holonic execution system is not forced to choose (analogy: having both a road map and a topographic map while using each map for which it is most suited).

Incommunicado

When discussing an open air engineering case, the issue of interrupted communication lines was raised. These applications often reside in remote areas where telecom infrastructure can be minimal. In the use cases that were considered, vehicles only communicate when they are within a short range (e.g., Wi-Fi or Bluetooth) and not at all at larger distances. Alternatively, long-range communication is highly constrained and/or overly expensive.

Conceptual solutions have been elaborated on paper (i.e., when the need would arise to address this issue of intermittent communication capabilities, the team knew how to tackle it and was confident the problem would be solvable). The explicit modeling of a corresponding reality proved useful. When subsystems reconnect, their internal models of the world have been disconnected from reality in various parts. Mutual updating of and agreeing on these models into a joint model for the connected systems involves accounting for the most recent information (i.e., reconnected holons get information from the source-of-truth holon) and the "shock" from this update to activities that have been working on disconnected (possibly outdated) models is a disturbance, which is handled by the mechanisms to handle change and disturbances that are present already. However, no in-depth research or extensive implementation was conducted.

Other Modeling Techniques

In intelligent traffic and transportation systems (ITTSs), complementary capabilities presented themselves (Philips et al., 2013). Traffic models – dynamic network loading (DNL) models – revealed to be capable of mirroring aspects that would be a challenging task for exploring and intention ants: congestions propagating backwards in a road infrastructure. Where traffic models have issues accounting for user intentions, DMAS has issues with collective behaviors that would require too many refresh cycles (iterations) to converge (assuming a straightforward DMAS design). Conversely, traffic models are highly effective and efficient at computing such collective properties.

In the MODUM project (modum-project.eu), this combination of D4U delegate MAS and models developed by the research community focusing on the application domain (ITTS) was investigated. D4U intention propagation predicts what the load/congestion levels of road segments will be; DNL models compute the effect of congestion propagation (backward) through the network. The intention propagation uses the DNL estimates for the virtual execution of their activity/traveling.

For such application-domain specific models to be available and effective, there has to exist a suitable stable target. Road infrastructure, modeled as networks of road segments, is present on a massive scale and it will be present for the foreseeable future. Practically, it is similar to physicists developing theoretical models about nature (e.g., Newton's laws).

In manufacturing or logistics, installations exhibit more variability both geographically (e.g., factories are designed differently) and time-wise (tomorrow's factories will be different). Models may be outdated by the time they are available or struggle to find sufficient users to justify the development efforts. A clever DMAS design, making a roundtrip (i.e., the exploring ant collects information on the way out and constructs a solution on the way back while using and needing this collected information), may be preferable in some cases. In other words, research still needs to establish where and how the above collaboration is possible and advisable.

Furthermore, a full-fledged holonic execution system, when deployed, is able to relax the demands on the models from the application domain experts. In the ITTS case, the execution system can be deployed to manage bus lanes going through bottlenecks (choking points) in the road network (note that they must not end somewhere inside the bottleneck). The execution system enables participants to use the remaining capacity, which is not used by the buses.

Here, the intention-propagation DMAS predicts congestion, and decision-making elements (intelligent agents) ensure that this never happens in reality (except in case of accidents/incidents, which is not considered further in this example). Travelers remain at home, in the office, etc. until capacity is available. Because of this service level, the DNL models must only allow to distinguish between congested and not congested; the execution system will even apply some safety margin to stay out of congested states. This relaxes the requirements on the DNL model considerably. Among others, it must not model nonlinear behaviors (e.g., waves of speed variation and the precise onset of waves), the impact of the type of crossing, etc. Also here, much remains to be investigated but disseminating how D4U creates novel opportunities for research and development may be needed even more.

Trust, Reputation, Commitment

In the MABE project (Reitbauer et al., 2005), the research expanded from closed systems (holonic manufacturing execution system (HMES) and HLES) into semi-open systems (Holonic Execution Systems for Networked Production – networked HMES). In networked production, access to the network is controlled but the members in the network are independent "selfish" organizations, participating in a positive sum game.

For such semi-open systems, it is not needed to make individual interactions "bulletproof" against malicious behavior. But it is necessary to have some "social control mechanism" in place. The network members are in a highly repetitive gaming situation where there is only a small amount at stake in each individual round or interaction (e.g., deliver one truckload) in comparison to the overall game (i.e., a long-term profitable collaboration and a company's reputation inside and outside the network).

The research activities developed a software framework, within an HMES design, to answer the challenge (Saint Germain et al., 2012). A framework was developed because research in trust and reputation remains in a flux and has no obvious solution to adopt and incorporate. A framework was developed to allow the use of trust- and reputation-handling mechanisms that are most suited to the situation as well as to switch when a superior solution presents itself in the future.

The framework allows to use a holon's track record (past behavior) in interactions. In particular, it allows to estimate what information provided by a holon actually means. When John communicates his intention to attend a meeting, tomorrow at 09h00, how likely is he to attend this meeting, at what time is he most likely to arrive, and how much uncertainty is there on this time estimate?

The framework, completed with suitable trust and reputation mechanisms, provides this service while enabling to account for context information. For instance, John's reputation concerning attending meetings professionally can be kept separated from his behavior outside his professional environment. Likewise, if John adds information about his commitment when promising to attend, the framework allows to account for this information.

Indeed, in this context, becoming explicit about commitment enters into the picture. Importantly, there exists no absolute commitment in a context of execution systems; reality does not provide locks, transactions in the manner of IT systems (databases). There always is the possibility of "force majeure," and execution systems have to cope with it (cf. Section "Software/System Development": the digital image of relevant reality must cover all possibilities).

In networked production, this framework implies that network members, exchanging information, have a social control mechanism that discovers and monitors the relationship between their words and their deeds. The mechanism makes no "moral" judgments. A member can be inherently unreliable. For instance, mounting the rotor of a wind turbine may take a long time when the weather causes delays. On the other hand, when John systematically books a cancelable flight at one airline while being waitlisted for his preferred choice, the first airline may become more prone to overbook John's flight.

The framework revealed to be useful outside its original target. When implementing the NEU protocol in real-life cases, some possibilities for (nice-to-have) enhancements emerged. For instance, the alternatives offered by the activity type may not be equal. Some activity steps on some resource instance will be mature, tried, and tested. Others may be untested and possible in theory only. The framework allows to manage this without having to change or enhance the NEU protocol in a significant fraction of this kind of situations. Likewise, it allows for straightforward enhancements of the NEU protocol (i.e., add some simple indicator), whereas the framework ensures a proper interpretation based on experience, track record, etc., including the adaptation of the interpretation when the corresponding reality evolves. Research in this matter still needs to be done.

Effort Versus Accuracy

The intention and exploring ants virtually execute their activity instances. This virtual execution involves decision-making procedures, which are performed thousands or millions of times before real-world activity step execution takes place. This is not an issue while these decision-making procedures remain computationally efficient (e.g., constant time or logarithmic). Real-world cases and moving into novel application domains reveal that there are many highly relevant situations in which this condition is not fulfilled.

To address this, the basic ARTI design using NEU protocols and DMAS needs refinement. There are virtual-execution components, which are used for DMAS execution, corresponding to real-world components that are used for actual execution. The virtual-execution components must be very efficient but are allowed to make mistakes. The impact of these mistakes must not be fundamentally different from other disturbances in the system. The real-world components must be accurate, however.

The following cases can be distinguished (at the present time):

- Intelligent beings model a corresponding reality. The accurate component versions normally use tracking (i.e., observe this corresponding reality) to achieve the required accuracy. The virtual execution components cannot wait for these observations and need a software model. This often will be straightforward and efficient (e.g., a time delay corresponding to the time to drill a hole).
- Regularly, some (machine) learning may be indicated to provide values for parameters and properties that are hard to estimate beforehand. Possibly, a computationally demanding model may be executed at a feasible frequency while computationally efficient models use its output to support virtual execution in a DMAS (e.g., perform inter-/extrapolation).
- Intelligent agents utilizing computationally simple methods in real-world execution (e.g., first-come, first-served) can be their own model for virtual execution.
- Intelligent agents utilizing computationally demanding methods (e.g., a 3D nesting algorithm) may need a computationally simple model for virtual execution, approximating the demanding one.
- A generic template for such a simple model is to execute the demanding method at a lower (feasible) frequency in combination with a simple mechanism to adjust the outcome of the demanding method in between executions of this demanding method. A machine learning solution is another possibility, especially when the detailed results of the demanding method are not needed for virtual execution (e.g., estimating what can be nested suffices).
- Humans take many of the real-world decisions and, often, need to "sign off" their decisions (e.g., doctors in the health care domain). The virtual execution counterpart may approximate this in numerous ways, depending on its use/responsibility. Typically, the authorized/responsible humans still make the decisions (per activity instance or type) once. And they control the autonomy settings of their virtual-execution components to be used in a DMAS.
- These autonomy settings determine how long and under what conditions the virtual-execution component may, for example, repeat the human's decision. In straightforward cases, a virtual-execution component

may take automated decisions. Often, the autonomy setting will steer a triage process. Some decisions are an automatic (e.g., what to do when medication was not taken at the prescribed time), some call and wait for a qualified human to decide, some provide a default action but inform a human to intervene at a time of his or her choosing, etc.

Of the above, the last case proved most interesting: how to include humans in holonic execution systems in full. This is discussed further in the next section.

TOWARD A HUMANE (MECHATRONIC) SOCIETY

Recent D4U research investigates and focuses on the humans within a holonic execution system, in particular, humans in integrated health care: patients, professional and layperson care providers. Moreover, the research has addressed open systems with very large numbers of users and resources. In particular, intelligent traffic and transportation presents these challenges. In combination, these investigations reveal that social aspects (e.g., social control) and empowerment of persons (e.g., concerning data about themselves) become most relevant.

It is necessary to address this social sciences and humanities matter for a successful introduction into society of the research developments discussed in this book. Conversely, introducing these developments into society creates opportunities to render tomorrow's society more humane. There certainly is a significant potential to create a more humane society when seen from today's prospects characterized by big data analytics in corporatist implementations running somewhere in the dark from an individual's perspective (Hildebrandt, 2015).

Social/Collective

D4U and its realization through ARTI and DMAS implementations differ from current prospects by their ability to incorporate intentions explicitly. These intentions are projected on a digital image of the world of interest, generating a prediction of the unexpected. More precisely, past experience is used solely to predict/model elementary activity steps such that future behaviors and states can be estimated also in situations where historical data on aggregate behaviors and states cannot be used.

This distinguishing feature has two implications. First, D4U can perform better simply because it uses more relevant information, which remains valid in more situations. Second, *D4U needs participation; by design, it cannot operate in the dark*. The latter is especially true when a D4U infrastructure is used to implement and achieve novel social interactions, which build upon a D4U digital image of reality that includes predictions.

In particular, persons – represented by software extensions acting in this digital image (e-Persons) – may interact socially in a virtual manner to manage their real-world activities (better). For starters, the D4U infrastructure eliminates or reduces the frustration from "if I had known this beforehand, I would have acted differently" where this "not knowing" originates from the inability to rely on the past to predict the effectiveness of choices concerning future actions.

For instance, commuters estimate – in a common mode – how a railway strike will affect congestion and decide massively to leave earlier or stay home. The result is a traffic jam almost an hour earlier than normal and less congestion than normal at the usual times. D4U would have witnessed this behavior virtually shortly after information about the railway strike became available. There would be ample time to observe the problems/congestions and missed opportunities. Commuters would get opportunities to adapt their intentions. Moreover, D4U infrastructure facilitates making commitments such that commuters will know their travel time beforehand with much less uncertainty.

Nonexhaustively, the following might be realized:

- Abolish "no good deed remains unpunished" by ensuring one's good deeds are accounted for virtually. This accounting needs mechanisms (legal or otherwise) to ensure commitment of all parties that are involved. For example, when a commuter leaves his home later, contributing to congestion avoidance during rush hour, he should not have (more) difficulties finding a preferred parking space as a consequence.
- Ensure virtually, with suitable commitment, that there is a fair deal before committing physically. Transactions, even complicated ones involving multiple participants, each performing an action, can be elaborated, detailed, refined, etc. before committing and its real-world execution. The "chicken game" can be played virtually. The collective may intervene if this game virtually ends in disaster.
- Virtual buildup to an agreement becomes possible. Participants may offer to contribute a small amount to show their goodwill but avoid offering larger amounts inviting exploitation and abuse (i.e., no signs of weakness). Virtual iterations may allow significant shifts, grasping win–win opportunities that are unreachable otherwise in the same amount of

time (because large steps toward the optimized solution, needed in slow real-world negotiations, would cause parties to consider the other to be weak and exploitable).

- Participative and collective change management is facilitated. Participants actively contribute to building a virtual image of a changed future (of an organization). In iterations and parallel versions, beneficiaries/winners and benefactors/losers become visible and can be accounted for. For instance, departments facing a reduced workload may be assigned novel responsibilities and training (where the persons involved are empowered to lead their reassignment). The change is executed as a transaction, reducing uncertainty and fear (avoiding and reducing resistance and sabotage).
- Organizations need less hierarchical control when the D4U infrastructure allows for coordination among peers by peers. Predictions allow adjusting activities over a longer time horizon and larger distances within organizations. Predictions allow higher management levels to intervene solely when indicated (in the predictions).
- Collectively and coordinated shifting of "what is considered normal" in a desirable direction when individuals volunteer small[11] contributions to the community concerned. These are visible because of the D4U infrastructure, inviting others to follow and join. The availability of a collective prediction creates possibilities to speed up what can be done compared to what real-world evolution could achieve.

Here, a D4U infrastructure (information and communication technology (ICT)) only offers a part of the solution. Legal support is needed (e.g., enabling to uphold commitments under pressure, to go against undesirable social pressure). Cultural innovation will be required to enjoy many of the potential benefits. For instance, a culture of discretion may contribute to self-management among peers, facilitating a frank communication where it is needed and beneficial.

Moreover, suitable initial targets need selecting and implementing. Unavoidably, there will be teething problems, which need implementation projects that are sufficiently small to succeed and large enough to learn. For instance, managing bus lanes – offering the free remaining capacity of the lanes to participating cars – by D4U will allow to mature its implementation before deploying it in other parts of traffic and transportation systems. Likewise, academic exercises, in which the D4U digital image is reused

[11]Small means "too small to exploit," a sign of good will, not a sign of weakness.

to simulate a corresponding reality, may provide insights and innovative designs. Also, some application domains may offer cases enjoying favorable conditions (e.g., high added value and relaxed conditions for D4U applicability). Overall, there remains plenty to research and develop.

Empowerment and Privacy by (Sociotechnical) Design

While applying the first part of this chapter to integrated care, the mirroring of human beings in a D4U digital image constituted a key concern. Every person, patient, as well as care provider (both professional and layperson) is to be extended virtually. For every person, there is an e-Person, which is an aggregate of the roles played by this person:

- A person as a valuable resource, which has finite capacities and capabilities. Typically, this will be an aggregated resource (e.g., acknowledging that a patient has organs, which are affected differently by medication, food intake, exercise, diseases).
- A person as a (composite) activity.
- A person as a decision maker concerning activities.
- A person as a decision maker concerning resources.

Overall, there will be an e-Person that is a relatively complex aggregate (holon) extending the real-world person in a D4U virtual reflection of the world of interest. This e-Person supports (at least) two modes or versions: signed-off and virtual-execution (cf. above). In all situations, the human remains in control, determining what is signed and how the virtual-execution version behaves. Moreover, this e-Person constitutes a starting point for information retrieval (by computer processes behaving like web crawlers).

This results, naturally, in a patient-centric design. All patient-related information is accessible through their e-Patient (a patient's e-Person). An e-Doctor (a doctor's e-Person) knows patients and retrieves information via their e-Patient. The doctor's data files are kept by the e-Patient, but parts might be encrypted such that only suitably authorized persons can access it. The e-Patient typically resides in a properly accredited data server ensuring minimal service levels while a patient is free to go beyond this. Note that commonly available technology allows to have the e-Patient available 24/7, 99.999% of the time. In other words, whenever there is a connection (Internet or another network), the e-Patient is available.

The above separates concerns:

- Suitably qualified ICT companies/organizations ensure that data are not lost, stolen, or corrupted. They ensure that the patient has no difficulties to comply with legal requirements. They offer additional services, allowing the patient to benefit from the latest technological possibilities.

- Care providers and organizations are able to manage information access by encryption to ensure it is understood correctly when used. This should maximize accessibility and control by the patient. The patient must even be able to manage accessibility and, e.g., decide about data availability for research purposes without requiring consent of care providers for most data.
- Unless there is a valid reason to do otherwise, even encrypted data must be accessible when the reader is qualified (i.e., understands it and is under an obligation to respect confidentiality). On the other hand, providers may have a fully private type of document, to avoid a defensive attitude about registering relevant information (e.g., about a patient's lack of hygiene).
- The patient is conceptually in control by default. The e-Person is the starting point for all data storage, manipulation, and retrieval. There is no corporation or administration that intervenes when data access is concerned except for encryption (to handle interests that the patient cannot manage or is not entitled to manage).

This patient-centric design will reduce and master complexity with a lot of ease when compared to prevailing schemes in which some external (corporatist) organization is, conceptually, the information repository and access controller. For example, when the patient's GP retrieves data through the e-Patient, relevant information can be volunteered (e.g., medication prescribed by a kidney specialist, a slimming food supplement taken on the patient's own initiative). This is D4U delivering SSOT.

Summarizing the above, a patient empowering design is the natural, perhaps only, manner to elaborate a D4U health care execution system for integrated care. Similarly, care providers will be empowered when they are in control of their e-Persons. However, the ICT alone is insufficient to empower. Policies need to be facilitated and even enforced. A culture of a person-empowering political correctness will help a lot.

Note that these e-Persons can be implemented as communicating computing processes that are always available (i.e., access to a computer, a computer network or Internet will be less available by an order of magnitude). In other words, there is no justification for retrieving information from a source other than the e-Person concerned. Thus, to empower it suffices to:

- Give every person maximal control over its e-Person(s).
- Make it a political/ethical/deontological incorrect action to access, store, or manipulate data without involving the e-Person as the entity in charge.

- Legally and technically support politically/ethically/deontologically correct behavior.
- Provide ICT service providers that ensure this high-quality storage (bank/notary-like).

This empowerment provides synergetic opportunities to ensure privacy. As data about a person are retrieved through their e-Person, the following examples can be implemented:

- *Please forget.* The e-Person, when asked to check some fact/data/information about its person, may indicate that this is marked to be forgotten. Similarly, the e-Person may refuse to confirm (i.e., qualify it as unknown). An ethical code may now require a justification to refuse the preference expressed by the e-Person. Technical and legal enforcement may enforce publication of such information as "unconfirmed", "marked as to be forgotten." Receivers of information may filter this kind of information out. Privacy advocates may spam readers looking for unconfirmed information.
- *Power-abusing parties seeking unauthorized access to information.* An extreme but nevertheless important aspect of privacy protection is nondisclosure of information to an unauthorized party while hiding the nondisclosure. With the e-Person as the SSOT, sensitive data will be refused to anonymous requesters. When the requester is powerful but unauthorized, the e-Person will honor the request but communicate information – of its person's choice – to which this party is entitled. It can be incomplete information or even modified information. When this unauthorized party requests the same information through another party (e.g., GP about a patient), this intermediary party will contact the e-Person concerned and forward whatever this e-Person provides (or information consistent with it).

The above illustrates how truly novel possibilities may present themselves. However, much still needs to be discovered and investigated before it can become a part of reality. Also, a lot can be designed and implemented poorly or wrongly. Looking into what the future may bring, recognizing problems, risks, as well as opportunities and benefits is needed, preferably beforehand rather than being forced to "repair" our society. Surprisingly, we may learn that supporting "bad behaviors" can be beneficial when constructing tomorrow's humane mechatronic society (cf. above: a collective ability to fabricate stories to cope with power-abusing parties making unauthorized requests).

Concluding Remarks

This section only scratches the surface of what still needs to be discovered. Only two *dimensions* have been recognized and presented above: social/collective behavior and human empowerment/privacy. There probably are others. And each of those needs further elaboration, refinement, correction, and prioritizing. The research discussed in this section clearly is subject of some exciting future research.

Note that D4U – including its ability to predict the unexpected – has a competitive edge over current developments such as big data analytics: *the best way to predict the future is to create the future*. D4U enables to create the future collectively and iteratively while it requires and empowers all persons involved to participate. It does not operate in the dark because it relies on the active and transparent participation of all persons and parties involved (unlike for instance big data analytics) and, as it keeps its options open, it are the participants who create their future collectively. Highly interdisciplinary research may discover how to do this, and to do it correctly. It may enable to address societal issues before they constitute a real-world problem by codesigning the ICT, and the legal, social, economic, educational, philosophical, and political elements.

SUMMARY

The extension to nonmanufacturing applications necessitated to revisit the reference architecture PROSA. In the resulting ARTI reference architecture, an adapted nomenclature for the underlying holons was adopted.

The NEU protocol is presented, describing the interaction between an activity instance and its activity type. It creates an independence – enabling a separation of concerns – between type holons (experts) and instance holons (managers). It further prevents exposure to the specifics of knowledge representation inside the holons.

New challenges such as interoperability, multi-resource allocation, stochastic issues, trust, etc. – arising when applying these results to real-world manufacturing or to new application domains – are described and potential research directions identified.

The authors express their conviction that D4U may pave the road toward a more humane (mechatronic) society, by stimulating social/collective behavior, by positive human empowerment while preserving citizens' privacy.

ABBREVIATIONS

ARTI Activity resource type instance
CPU Central processing unit
ITP Intelligent transport system
MVP Minimally viable product
NEU Next execute update
SSOT Single source of truth
DNL Dynamic network loading (in traffic models)

REFERENCES

Hildebrandt, M., 2015. Smart Technologies and the End(s) of Law. Edward Elgar Publishing, ISBN: 978-1-84980-876-7.

Huang, H., Task learning and execution for behaviour-based mobile manipulation. PhD thesis, KU Leuven, 2011. ISBN 987-94-6018-381-2.

Kruth, J.P., Detand, J., 1992. A CAPP system for nonlinear process plans. CIRP Ann. Manuf. Tech. 41 (1), 489–492.

Philips, J., Holonic task execution control of multi-mobile-robot systems. PhD thesis, KU Leuven, 2012.

Philips, J., Saint Germain, B., Van Belle, J., Valckenaers, P., 2013. Traffic radar: a holonic traffic coordination system using PROSA + + and D-MAS in industrial applications of holonic and multi-agent systems. Lect. Notes Comput.r Sci. 8062, 163–174.

Reitbauer, A., Battino, A., Saint Germain, B., Karageorgos, A., Mehandjiev, N., Valckenaers, P., 2005. The MaBE middleware in emerging solutions for future manufacturing systems. IFIP 159, 53–60.

Saint Germain, B., Valckenaers, P., Van Belle, J., Verstraete, P., Van Brussel, H., 2012. Incorporating trust in networked production systems. J. Intell. Manuf. 23 (6), 2635–2646.

CHAPTER SEVEN

Case Studies and Research Projects

Paul Valckenaers*, Hendrik Van Brussel**
*Faculty of Engineering Technology, KU Leuven
**Faculty of Engineering Science, KU Leuven

This chapter discusses case studies and projects applying and investigating design for the unexpected. They focus on the design and development of holonic execution systems in a wide variety of manufacturing and nonmanufacturing applications. This variety of cases shows the universal applicability of the product-resource-order-staff architecture (PROSA) / ARTI delegate multiagent system (DMAS) framework. The obtained results emerged from a synergy between more fundamental research results and the application of these in real-world cases. The case studies and research milestones are presented in chronological order respectively for manufacturing and other application domains. The adoption of a chronology in ordering the different cases reflects the evolution in the minds of the authors to come to the final results captured in this book.

Execution systems manage operations in real time. They only trigger system-specific activities, leaving detailed control to other systems. For instance, a manufacturing execution system (MES) leaves the pick-and-place of a dashboard into a car body to the robot controller and/or human workers. In contrast, the MES manages the routings and processing sequences for these products (e.g., car bodies, doors, and dashboards).

The primary concern is to cope with all relevant aspects of a world of interest (e.g., manufacturing equipment and products), including – in view of this book's title – unexpected aspects. Note that this primary concern is irreconcilable with performance optimization, imposing restrictive views on the world of interest. Because the case studies invariably have combinatorial solution/search spaces, the presently known optimization techniques can only be connected in an advisory role to the world of interest.

In this respect, it is important whether the execution system is, for instance, able to represent and exploit all the possible routing and processing alternatives for products (i.e., this is desirable). In contrast, rendering a

Design for the Unexpected. http://dx.doi.org/10.1016/B978-0-12-803662-4.00007-2

routing or processing alternative unreachable, typically to accommodate an optimization procedure, is a violation of the design principles (i.e., this is undesirable).

Historically, holonic MESs have been the subject of our research. However, the scope of the case studies is broader. It includes the application of product-resource-order-staff architecture (PROSA) and delegate multiagent system (DMAS), developed for manufacturing, to applications outside the manufacturing environment, notably in its consolidated ARTI version. In general terms, *design for the unexpected* (D4U) can be applied successfully whenever:

- Activities are executed on resources.
- Activities are subject to constraints that render myopic[1] decision making ineffective. Typically, decisions concerning these activities have effects later and elsewhere, and future activity interactions affect performance significantly (e.g., create congestion).
- Virtual execution of these activities in a digital mirror image of the world-of-interest is possible and can be much faster than in reality. This allows, among others, exploring and intention ants to "predict the unexpected."
- The socio-economic benefits of such an enhanced coordination more than compensate the (recurrent) cost and effort needed to maintain and operate the coordinating execution system.

MANUFACTURING CASE STUDIES

The PROSA Precursor: FACCS (1985–1990) (Valckenaers et al., 1995)

FACCS, the flexible assembly cell control system, is a holonic MES for a robotic flexible assembly cell. This cell comprises four robotic assembly stations and a transport system allowing product carriers to visit these stations in any sequence (Figure 7.1). This flexible assembly system was part of a research setup in the research lab of the authors at KU Leuven.

In view of 2015's state of the art, FACCS has the following contributions:

- Product-driven manufacturing control. Each production order self-manages its routing and processing step sequencing. This results in a (robust) heterarchical control mode.

[1]Only looking at past and present states.

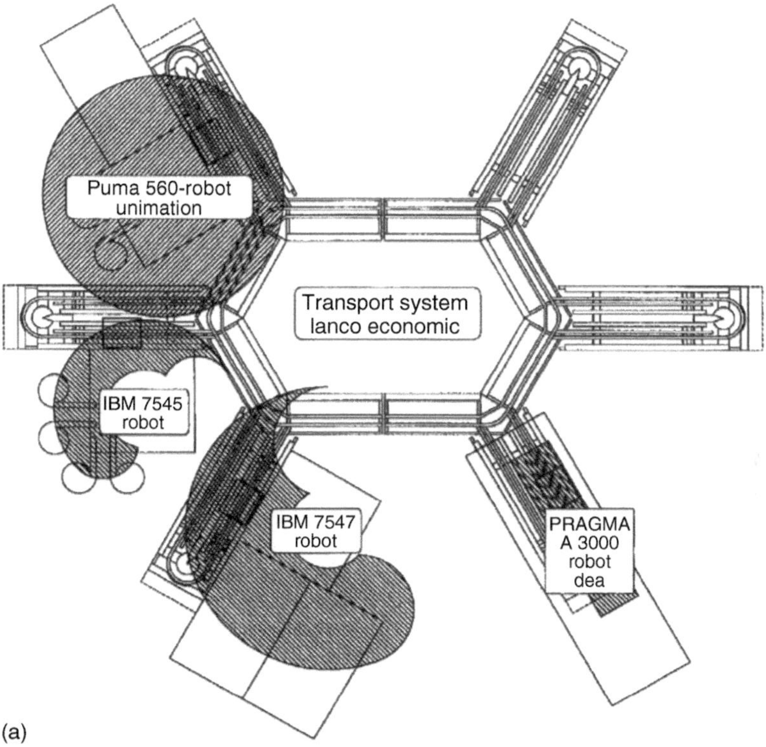

(a)

(b)

Figure 7.1 ***Graphic representation (a) and photograph (b) of the holonic flexible assembly cell at KU Leuven.***

- Process plans are based on *dynamic precedence graphs*. These graphs are able to represent alternative processing sequences, including mutually exclusive sequences, handle result-dependent sequences (e.g., repair when test results indicate the need), and allow simultaneous execution of processes for the same production order (i.e., lock/race-free updating of the state of a graph).
- Uses alternative processing sequences in opportunistic routing and/or sequencing to handle disturbances in a heterarchical control mode.
- Is able to cooperate hierarchically with a reactive scheduler to optimize performance and combine this with the heterarchical mode to cope with disturbances and changes.
- Supports a fine granularity for the primitive processing steps used in the dynamic precedence graph. The requirement limiting this granularity is that the products are transportable and storable before and after executing such a primitive processing step.
- Is able to produce any combination of the available primitive processes in a batch-of-one mode.
- Connects production equipment to the control system through device drivers, thus avoiding exposure for the holonic MES to device-specific programming languages, tools, or environments. This complies with the autocatalytic set insight (cf. Chapter 4). Note: this device driver approach has been successfully deployed in an industrial robot application as a spin-off activity.
- It is process-agnostic, which implies that the control system is able to manage any mix of production processes (assembly, machining, painting, packaging, etc.).

Relative to 2015's state of the art, FACCS has the following limitations:

- Job shop assumption. The self-managing production orders assume that they can loop around the production system and reach workstations that are offering primitive processing steps without having to plan or reason. For example, FACCS is unable to handle a flow layout in which a needed processing capability may be passed and become unreachable (or impractical to reach).
- Each workstation is modeled as sets of IDs corresponding to the primitive processing steps that the workstation is capable of performing. For example, there is no model of tooling, fixtures, operators, etc.
- Process plans have a predefined data format based on the IDs of available primitive processing steps. For example, any information depending on the product model needs to be retrieved by the implementation of the primitive processing steps (using the order ID as a key for data

retrieval). Such information remains invisible to and cannot be used in the process plan.

- The heterarchical mode is myopic. The self-managing production orders behave opportunistically without assessing future repercussions or impact on other orders.
- Reusability of the software and system components is low/questionable. Dynamic precedence graphs are an option for product holon implementations but must be used internally and cannot be exposed to, for example, the order holons (cf. NEU protocol). The device drivers and primitive processing step implementations are case-specific. The heterarchical control mode under the job shop assumption is easily reimplemented, while relaxing this assumption requires a significant effort (cf. the other case studies).

Relative to 2015's state of the art, FACCS utilizes the following information and communication technology (ICT):

- FACCS is programmed in mainstream object-oriented programming technologies (C++, Java) on mainstream computer platforms (Windows, Linux). This complies with the autocatalytic set guideline/insight.
- FACCS has no simulation capability.
- FACCS is connected to industrial equipment through industrial automation technology (i.e., PLC for the transport system and robot controllers for the assembly robots). However, FACCS minimizes its exposure to these industrial technologies as they lack critical user mass (i.e., they are violating the autocatalytic set guideline/insight). The industrial controllers are used to program primitive services, which are made available on the mainstream computer platforms in mainstream programming languages as device drivers. In case of machine tools, a DNC option would be desirable, allowing to make the machining functionality available on the mainstream platform.

The First Milestone: PROSA (1990–2000) (Van Brussel et al., 1998; Simon, 1990)

Continued fundamental research on the FACCS flexible assembly cell resulted in the PROSA reference architecture. This reference architecture respects and accounts for the fundamental insights, discussed earlier in this book. Consequently, PROSA represents to a large extent what is unavoidable, and it discourages imposing of arbitrary design choices.

In order to achieve this, PROSA is only a reference architecture; it is not a system architecture or an implementation. The research team stopped elaborating this particular research result as soon as further development required the introduction of arbitrary design choices. Such choices had to be made for the implementation of research prototypes but they never became part of this reference architecture.

In view of 2015's state of the art, PROSA has the following contributions:

- The reference architecture provides a terminology, greatly facilitating the communication and discussion in the domain. Chapter 5 discusses this in detail.
- PROSA cleanly separates the structural aspects from the control aspects. PROSA emphasizes the structural aspects as the foundation for holonic MES development; the control aspects follow in later design and development phases. Note that this ensures scalability as systems mirror the structure of the manufacturing system and activities concerned.
- PROSA aggregation – which can be time-variant – corresponds to the insights from Herbert Simon (Simon, 1990), which constitute the core insights in holonic systems.
- PROSA specialization and aggregation enable reuse and increase the achievable user mass for software components.
- PROSA separation of technical concerns (addressed in product holons interacting with resource holons) from logistical concerns (addressed in order holons interacting with resource holons) contributes significantly to achieving critical user mass and software reusability.
- Product-driven manufacturing control. PROSA mirrors a reality in which production activities execute on manufacturing resources, and PROSA considers both to be equal (i.e., they are so-called first-class citizens). Industrial practice often collocates all intelligence with the resources whereas the products only have data sheets/forms attached. This preserves the nonautomated practice in which humans operate equipment whereas products do not have a human companion. This industrial practice violates the insights and guidelines of design for the unexpected when intelligent machines have to rely on expectations on the manner in which they will be used. In PROSA, intelligent machines – resource holons – only have to account for what they are and what they are capable of. How these machines will be used is known by product-related holons. Among others, this increases configurability, reconfigurability, software reusability, and the ability to achieve and maintain critical user mass.
- PROSA imposes minimal requirements on the nature of the system that is to be controlled.

Relative to 2015's state of the art, PROSA has the following limitations:

- PROSA is a reference architecture, which means generic rather than specific. As a consequence, its ability to handle concrete multiple production system types – job shops, flow shops, production lines, etc. – has been achieved by leaving most of the development work to be done.
- The base holons, without schedule-generating staff holons, are myopic. Intelligent products will be navigating and shopping for resource services, which perform the processing steps, like car traffic in which drivers navigate to get serviced (work, school, home, shop). There is no support for the base holons (product, resource, order) to anticipate the consequences of their decisions and choices at later points in time and/or remote location in the factory. Nonetheless, cooperation with planners is covered through staff holons.
- PROSA separates the product-related technical and logistical aspects (in, respectively, product holon and order holon). PROSA neglects to do the same for resource-related aspects. Here, PROSA mirrors its source of inspiration – computer operating system kernels – which also fails to separate resource-related aspects (i.e., resources correspond to CPU cores, orders correspond to data segments of threads, products correspond to code segments of programs). ARTI refines PROSA in this respect.
- Research has addressed the development of reusable software that implements PROSA. This research has not produced software results that have survived until today. Partially, this was caused by a lack of insight and know-how. Partially, this was caused by subsequent research discovering important functionality that could be added to the PROSA foundation without violating the design for the unexpected insights and guidelines. Partially and most decisively, it was a consequence of mainstream ICT limitations. Even when developing the right software components, the effort needed to develop, maintain, and support would have remained too high.
- The software developments have only a limited capability to simulate faster than real time (i.e., speed-up is a fixed factor). Such a speed-up corresponds to lowering the processing speed of the computers on which the MES executes.

Relative to 2015's state of the art, PROSA utilizes the following ICT:

- PROSA-related software development uses mainstream OO-programming technologies (C++, Java) on mainstream computer platforms (Windows, Linux).
- PROSA-related research developed a real-time simulation capability with software in the loop. The factory simulation – without the holonic

MES – was done in ARENA™. It was used in real-time mode and connected to the MES through a TCP/IP socket. The MES could not distinguish being connected to the simulation from being connected to a real production system.

The Second Milestone (Part 1): MASCADA and Food Foraging in Ant Colonies (Peeters et al., 2001)

The myopia of a straightforward PROSA design remained a major concern. Scheduling staff holons may only give advice for good reasons: state-of-the-art planning and scheduling technology fails to comply with *design for the unexpected*. Furthermore, this situation is unlikely to improve as execution systems have combinatorial solution/search spaces. Optimizing planners and schedulers have to reduce this space to a polynomial one. As a consequence, developing a definitive and optimizing solution to this challenge amounts to breaking encryption systems that are currently used (cf. NP completeness).

In other words, planners and schedulers have to make choices – arbitrary to a significant extent – to produce their results, which inherently conflicts with design for the unexpected. As the research team aims for a holonic execution system foundation complying with design for the unexpected, scheduling holons had to remain optional staff holons. Apparently, eliminating myopia in PROSA-based execution systems was asking for a more innovative solution.

Stigmergic Coordination and Control

The answer was discovered in two steps. First, heterarchical control designs in which resource holons will signal to order holons when they are about to finish their current task were investigated by other research teams. This was, however, too simplistic to remedy this myopia. Such an order attraction signal should reach prospective order holons after being enriched with information accounting for space and time. It should account for a predicted workload of the resource holon (a timetable). And, it should account for the intended journey of the workpiece to the resource (estimated order arrival time in this timetable). Moreover, there should be indications of valid routing options – road signs indicating which processing options are available/installed in which directions – to prevent order holons from selecting routings that lack the required processing capabilities.

Second, the food foraging coordination mechanisms used in ant colonies revealed how information may be deposited locally in an environment – referred to as stigmergy – to inform members of the colony about a remote fact. When an ant discovers a food source, which is too large for a single ant to carry, it will deposit a chemical, called a pheromone, on its way back to the nest. Other ants will then follow this pheromone trail to the food source.

Two properties of this stigmergic coordination are interesting. First, pheromones evaporate. When an ant follows the trail and brings (some of the) food back to the nest, this ant deposits more pheromone. When the food source expires, the ants start exploring for food and will not return to the nest. The trail is no longer reinforced; the pheromone evaporates and disappears. This mechanism allows the ant colony to adapt and cope with the dynamics of its environment. Second, the ants use their environment as part of their solution. Ants have no model of their world inside their brain. The pheromones are deposited on the real world. Whether their world is (geometrically) simple or complicated has no impact on the ants' coordination mechanism.

However, ant colony coordination is limited to the past and present. The MASCADA project transformed this stigmergic coordination to cover the future as well (i.e., local information about remote facts in time and space). To this end, the MES uses a software environment in which production activities execute virtually. During such a virtual execution, information can be collected, generated, and deposited (i.e., digital stigmergy). In addition, this information disappears unless it is regenerated to cope with change and disturbances. In other words, the ant colony solution is applied in a software that mirrors the world of interest, allowing to try and evaluate (a lot of) alternative courses of action before committing to the most desirable course of action that was encountered in this virtual world.

Car Body Painting – Holonic MES for Flexible-Flow Shops

The MASCADA project developed its solutions while addressing an industrial case. A car body paint shop was an excellent case to test the applicability of an HMES for flexible *flow shops*. Each day, this large shop, comprising six floors, paints more than 1000 car bodies, of different types and in different colors. It comprises more than 400 manufacturing resources: unidirectional and bidirectional conveyors, turning tables, lifts, painting boots, etc. (Figure 7.2).

These resources are arranged in a complex topology, in which loops are present. The system has built-in redundancy; that is, for each processing step, multiple resources can be chosen. Similarly for the transportation, more than one routing option is available to move a car body from one processing unit to the next. As the result of a production step is uncertain, the next processing step for a car body will depend on the outcome of the previous one. This means that it is sometimes necessary that a product makes a loop

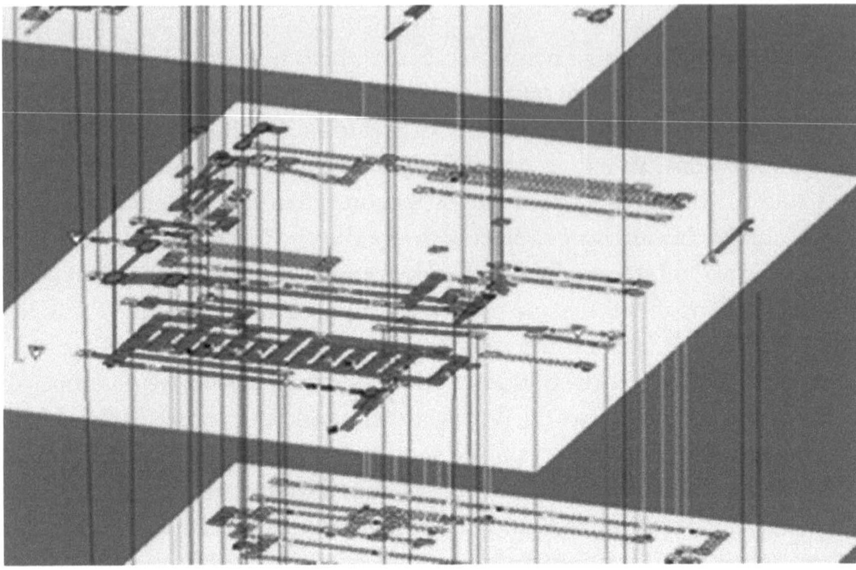

Figure 7.2 ***Partial view of the six-story spray painting shop simulation for car bodies.***

through the paint shop. The main performance measure in this paint shop is throughput. The throughput can be influenced by the batch size. Throughput losses are caused by color breakdowns on the painting lines and blockages on the transportation system.

The control system, a precursor DMAS system, is responsible for the routing of the car bodies through the paint shop, and it has to maintain the required throughput in the face of disruptions. Because of loops in the transport system, the control system also has to deal with deadlocks. Therefore, the intelligent products (corresponding to the car bodies) use a layered decision mechanism to choose their next processing step.

The first control layer addresses feasibility. This layer is responsible for deadlock avoidance and ensures, for instance, that a car body is not transported in a direction that lacks the necessary processing capabilities. The second layer is an optimizing layer, doing things such as satisfying production goals (e.g., maximizing throughput or respecting due dates), optimizing bottleneck usage, and avoiding material flow jams upstream and downstream. A third layer tunes online the parameters of the optimizing functions of level 2. All these layers are application specific (plug-ins) and can easily be replaced if necessary.

The control system is also responsible for batching the car bodies for the painting process. Small batch sizes lead to more setups leading to lower throughput. Moreover, as batches are small, there are more defects, and more car bodies have to be repainted, lowering the throughput even more. To deal with this issue, the intelligent resources corresponding to the painting equipment propagate information about their planned batches (size, color, time window, etc.). The intelligent products use this information to decide whether or not to join a certain batch.

MASCADA – Contributions and Limitations

In view of 2015's state of the art, MASCADA made the following contributions:

- MASCADA delivered the innovation needed to address myopia in PROSA while respecting the insights and guidelines of design for the unexpected. It invented the use of stigmergy in a software environment supporting virtual execution as needed by the coordination and control.
- MASCADA addressed a full-scale industrial test case (i.e., a large car body painting plant comprising more than four hundred pieces of equipment and over a thousand products being processed at the same time).
- MASCADA is fully compliant with PROSA. At no point was there a necessity or benefit to be gained from deviating from the reference architecture. Or, in other words, the research did not encounter a situation in which PROSA would benefit from a modification or revision.
- MASCADA developed the first generation of DMAS. More precisely, the following was developed for the industrial case study:
    - Feasibility ants. These digital ants virtually traverse the flexible-flow shop in the reverse direction of the products. They collect information about the processing capabilities of the resource holons, and they deposit this information at routing points such that order holons may select valid routings only.
    - Order holon ants. These digital ants are created by order holons and virtually execute the remaining processing and transport steps for their order (i.e., a car body). During this virtual execution, the ants inform the affected resource holons about their intended visit. This allows the resource holons to construct a timetable, which predicts their future workload.
    - Resource holon ants. These digital ants take the timetable of their resource holon and virtually traverse the flexible-flow shop in the reverse direction of the products. The ants collect information about the travel time toward their resource. At routing points, the ants deposit a timetable for their resource adjusted for the accumulated travel time. In other words, order holons at a routing point will observe timetables indicating processing capabilities and (free/available)

capacities adjusted for the time needed to reach a resource. Specific for the industrial case, order holons also see whether a painting workstation needs to change color when its car body will arrive.
  - A number of case-specific stigmergic mechanisms have been developed. This mainly demonstrated the contribution a stigmergic information infrastructure can make toward the implementation of case-specific mechanisms (i.e., fast and with little effort). An example is batch-building mechanisms, aiming to avoid color changeovers at the painting workstations.
- A deadlock avoidance mechanism was developed for the industrial test case.

Relative to 2015's state of the art, the MASCADA results have the following limitations:

- The resource holon ants are overly specific and their design struggles to remain usable beyond flexible-flow shops, as encountered in the car body painting case. In particular, the collection and computation of the accumulated travel time for an order toward a resource is problematic. Subsequent research developed a solution that is both simpler and widely applicable. The MASCADA resource holon DMAS has been renamed into the resource intentions DMAS. In some systems, it is still used but solely in a supporting role where it is possible to stop propagation as soon as computing the accumulated travel time becomes problematic.
- The design critically depended on the correctness of the feasibility holon information and the deadlock prevention. The above-mentioned replacement for the resource holon ants has relaxed this requirement significantly.
- The order holon ants fail to behave in a socially acceptable manner. Order holon ants virtually execute to inform resources about likely future visits; this involved a randomized decision-making mechanism. However, order holons had no mechanism to "stick to their intentions," and the randomized decision-making mechanism was executed, both during refresh and for the actual routing selection, without accounting for earlier propagation of order intentions. This severely compromised the prediction accuracy of the resource timetables. The main reason for this shortcoming was lack of experience for the junior researcher and lack of time for the senior researcher (attending to an overzealous project review board focusing on scheduling and optimization aspects). Subsequent research did remedy this (Hadeli 2006).
- MASCADA did not develop software that survived the project. First, the above two limitations, and the subsequently developed solutions,

would have rendered much of the software obsolete. Second, the available software technology was inadequate in view of the holonic MES still remaining a research prototype (which denied access to human and financial resources that would have been required).

Relative to 2015's state of the art, MASCADA utilizes the following ICT:

- Software development in Java technology on Windows.
- Real-time simulation in ARENA™ with MES software connected via a socket.

The Second Milestone (Part 2): Photographic Foil Facility (Saint Germain and Verstraete, 2002; Van Belle et al., 2014)

Where MASCADA breached the wall of decision myopia without the introduction of arbitrary design choices, the photographic foil case delivered the currently surviving design. As can be expected, industry offered a set of manufacturing execution challenges, explained hereunder, that could not be answered by the available commercial solutions, not even after customization and not even partially.

The photographic foil facility produces photographic products out of large rolls of photographic foil. A customer order consists either of a stack of sheets or a roll of a given kind of photographic paper. These rolls and sheets are made out of big rolls, called master rolls, by dividing these big rolls into smaller pieces. This way, these master rolls can be associated with multiple customer orders, which only require some part of the big roll. Conversely, a customer order can be associated with multiple master rolls since the units in a single customer order can specify different types of foil (each master roll only contains one type of foil).

In other words, there is a many-to-many relationship between master rolls and customer orders in the system. To form a product, the master roll has to be split first lengthwise, by a process called slitting. This operation divides the master roll into pieces, called reels. The second operation splits the reels along the width of the original master roll. Depending on the products to be made, the operation is either cutting or rewinding. The final product is a stack of sheets or a small roll, respectively. Figure 7.3 shows the elements of the production plant and their interconnection.

PROSA+DMAS – The Definitive Design: Simple, Efficient and Robust

While applying and implementing the MASCADA solution to the photographic foil case, the team started with the order holon DMAS, which

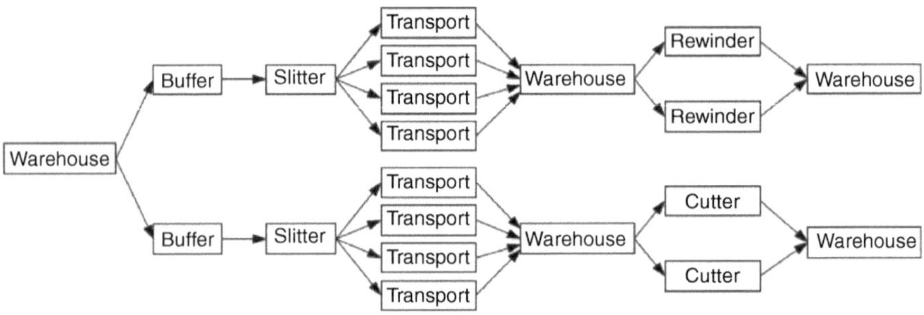

Figure 7.3 ***Photographic foil production plant.***

virtually executes the remaining processing and transportation steps for the order. The team realized that this DMAS functionality could be used for two purposes, requiring a minimal amount of software development for the required adaptation to one of the roles.

First, virtual execution can be used to inform resource holons about the intended visits by the order. This functionality was baptized the *intention propagation DMAS*, comprising a steady stream of intention ants. Second, this virtual execution capability was used to discover and assess possible routings and processing sequences. This was baptized the *exploration DMAS*, comprising a steady stream of exploring ants performing a decentralized and randomized search.

The difference between the two DMASs is that the *exploring ants virtually execute a possible solution*, selected by randomized decision-making mechanisms (see Chapter 5), whereas the *intention ants virtually execute the currently selected routing and processing sequence* (i.e., the order intention). The selection of the order intention is done by a given selection mechanism plug-in.

Importantly, the above selection mechanisms are easy to replace. The PROSA+DMAS design imposes no restrictions except that they perform the required selection and that they are efficient (i.e., they allow virtual execution to be faster than reality). In accordance with design for the unexpected, decision-making mechanisms must not build up inertia; it must be easy and fast to change and/or replace them.

Photographic Foil Facility – Contributions and Limitations

In view of 2015's state of the art, PROSA+DMAS has the following contributions:

- Lasting elegance. The MASCADA discovery, cracking the myopia issue, was translated in a superior design that is simple, intuitive, and widely applicable. It fully complies with design for the unexpected when it maximally solves the problems by mirroring the world of interest. Virtual execution of single steps constitutes its basis on top of which generic and generally applicable mechanisms provide the short-term predictions. These predictions are updated regularly. In the created information infrastructure, which contains these predictions, a decentralized search is facilitated. Indeed, this PROSA+DMAS design is minimalistic yet broadly applicable.

- PROSA+DMAS coordination is (more) robust toward deadlock. It generates short-term forecasts that will predict/see an imminent deadlock situation. Close to deadlock, only a few activities remain unblocked. The assessment of solutions, discovered by exploring ants, will see no benefit in being eager/early. It suffices to reward late commitment and/or penalize useless work-in-progress when selecting the order intentions. This will discourage the order holons from jamming up the manufacturing system. It may still be necessary to provide deadlock handling in an execution system but its importance is reduced significantly.
- PROSA+DMAS reduces the requirements for the feasibility DMAS. When an exploring ant selects an unfeasible route, its virtual execution fails to finish the product. This translates in an extremely low score, and it will never become the intention of its order holon. In other words, it suffices that an adequate number of exploring ants selects a feasible route. Wasting some effort on exploring infeasible routes is a small price to pay when this allows using a simple feasibility DMAS and/or circumvents the need to proof that this feasibility D-MAS is "watertight".
- As was discovered subsequently, sticking this close to reality rendered the design highly suitable for further developments. It enabled cooperation with scheduling staff holons, sticking to intentions, and adding a trust framework. In a way, it confirmed design for the unexpected when the design revealed to be able to cope with future – unexpected – demands.
- The industrial member of the research team – who was managing the real photographic foil facility – noticed that the development is multipurpose:
  - It can be used as a holonic MES for the actual production system.
  - It can be used as a simulation of a production system connected to its holonic MES. Both production system and MES can be subject of experimentation to optimize decisions on what system to build, how to operate it, etc.
  - The holonic MES, while connected to the real system, could be "forked" and start predicting two or more alternatives. This fork starts at some point in the future where the predicted performance

will become known. At that point, the best-performing alternative is adopted.

- The software development replaced the commercial simulation software (ARENA) with a pure Java emulation connected to the holonic MES, which was also written in Java. First steps were taken to have a hybrid simulation (real-time and discrete event) as well as tools to analyze the results of simulation runs (visualization during the simulation run as well as afterward using the log files).

Relative to 2015's state of the art, the photographic foil results have the following limitations:

- The case study did not develop software that survived the project. Nevertheless, it constituted the basis for all software development until 2010. From 2011 onward, development in Java was phased out.

Relative to 2015's state of the art, photographic foil case utilizes the following ICT:

- Software development in Java technology on Windows.
- Emulation/simulation in Java. Originally using a simulation library in Java but, finally, simulation was performed in self-written Java code.

Modular Plant Architecture II – Machine Shop (Zamfirescu et al., 2003)

Around the same time as the photographic foil development, a machine tool shop was addressed in the Modular Plant Architecture (MPA) project. Both projects shared solutions and software development. This project focused on the industrial test case while benefiting from the past (PROSA, MASCADA) and the present (photographic foil).

The case study concerns a job-shop used for the manufacturing of long weaving machine components (Figure 7.4). The production is organized around an automatic storage and retrieval system (ASRS) that acts as a temporary buffer for pallets shaped as containers. A rail-based transport system,

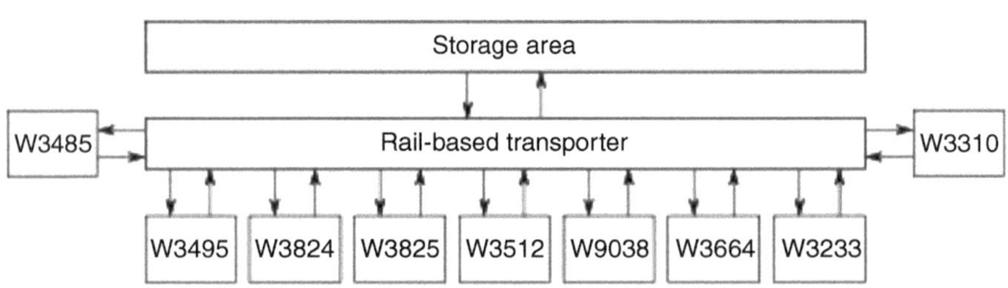

Figure 7.4 ***Schematic of the machine shop.***

called "tram" in the factory jargon, is used for storing/retrieving the containers into/from the ASRS and changing the containers at the workstations. Each container contains a variable number of identical parts traveling together till the completion of their processing plan. The machines are grouped in workstations, with a variable number of container docks and with different processing capacity. Typically, a workstation holds two containers: an empty container to be filled with the finished parts and a full container with parts to be worked on.

Inside the workstation, a part is taken by the machine operator from the full container and loaded into the processing machines, processed, and then unloaded and stored in the originally empty container. When this last container is full, the ASRS is prompted to take it away. Because the tram has two container docks, prior to picking up the finished containers, it travels to the ASRS to bring the next container that is going to be processed in the requesting workstation. Therefore, once the tram took the container with the finished parts, it unloads the next container without an additional movement. Finished parts are stored in the ASRS and retrieved in a given number on a daily basis, according to the assembly orders. There are also some trolleys that can take over the transportation effort as required. The machine operators are assigned to workstations and not to a single machine on the basis of their skills, shifts, and preferences. Overall, the plant holds the characteristics of a classical open job-shop, with different alternatives to carry out an operation, either processing or transportation.

Relative to the above-discussed developments, MPA-II made the following contributions:

- During the MPA project, emulation technology was elaborated in Java, replacing the commercial simulation package. The main advantages of this change were:
  - Training. Developers no longer had to master two programming technologies. They only needed to use Java. They no longer needed to know how to build ARENA models and templates. They only needed to acquire programming skills representing a universal/multipurpose value to themselves, and no longer did they have to spend time and effort on mastering an antiquated niche technology.
  - Hybrid simulation/emulation. The holonic MES and the Java factory emulation were cooperating closely. In particular, the MES is able to signal when all (emulation) events have been processed, which allows the emulation to jump in time to the next event. This speeds up simulation runs. Conversely, while the MES is processing events, the emulation executes in real-time to realistically account for the fact

that the world will not wait while the MES is "thinking." The MES also uses this to generate its internal time-based events (e.g., trigger the refresh of a digital pheromone), which then may benefit from the speed-up as well.
  - Support for analyzing the results, both during a simulation run and afterward through off-line data processing.
  - This technology survived the project in our own research environment.
- Demonstration of the capability to address an industrial case, possessing specific features to optimize production, and to contribute to resolving real-world issues. For instance, transport units contain about ten product parts, the transporter carries two containers, etc., which would not fit naïve models of a world of interest (e.g., the nominal job shop model). Note that technology developed for flow shops was used for a job shop without the need to start over.
- Likewise, the prediction-generating capabilities are instrumental for the effectiveness and comfort of the human workers, who in the existing machine shop faced uncertainty about the arrival of work and/or tooling. Having a prediction enables them to coordinate their main task (machining) with the other task (e.g., maintenance).

Relative to 2015's state of the art, the MPA results have the following limitations:
- Software development still is too time- and effort-consuming for industrial deployment while the technology has to prove itself (i.e., if it could mobilize resources such as established ERP, it would be more than viable).
- Order holons fail to interact as soon as they are known (i.e., as soon as the HMES is informed about the corresponding real-world orders). The holons only start exploring and propagating intentions when they are launched on the factory floor. This disturbs the generated predictions, because their intention propagation affects operations in the immediate future, and the system needs some time to recover. When order holons are created some time before their order will be launched on the factory floor, their intention propagation only affects operations after their launch date/time, which may then allow accommodating them without disturbing actual operations. Project funding stopped before this could be remedied and experiments could be repeated.

Relative to 2015's state of the art, MPA utilizes the following ICT:
- Software development in Java technology on Windows.
- Emulation/simulation in Java. Originally using a simulation library in Java but, finally, simulation was performed in self-written Java code. Advanced

features include hybrid emulation (discrete event jumping ahead in time combined with real-time generating event while the MES "thinks").

MABE Project – Networked Manufacturing/Heat Treatment (Saint Germain et al., 2012; Saint Germain et al., 2011)

A last manufacturing case study addresses a highly automated heat treatment multiplant facility. This facility performs heat treatment of metallic materials and includes several processes: case hardening, vacuum hardening, induction heating, etc. The products demand a certain temperature trajectory inside the furnaces in order to reach the required quality. The time between different processes (e.g., between case hardening and tempering) should not be too long for some products. The various furnaces differ from each other in the range of working temperature and environmental condition (e.g., carbon level). The facility is organized as a job shop in which the baskets containing the metallic parts are transported automatically. Figure 7.5 shows the temperature profile and corresponding resources of the case hardening process.

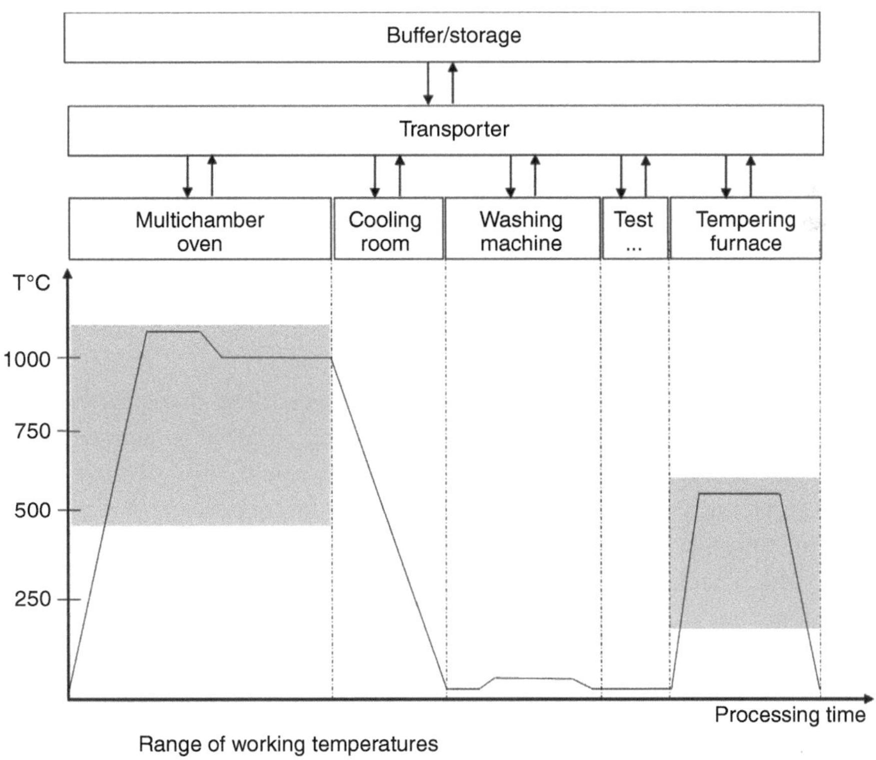

Figure 7.5 ***Case hardening facility.***

The control system is responsible for the routing of the to-be-treated metallic parts through the facility, ensuring that these parts receive a correct treatment. The intelligent resources correspond to the transportation and heat treatment equipment (e.g., furnaces, washing stations, and cooling beds). The services offered by these resources are used by the intelligent products, corresponding to metallic parts that have to be treated.

Specific for this application is that parts with compatible process temperature trajectories and environmental conditions can be batched. This batching, when properly executed, has a significant impact on the performance of this capital-intensive production system. Indeed, a fully loaded furnace and a partially loaded one operate almost at identical cost, whereas the output differs significantly. The intelligent products can make use of a delegate MAS to discover batching opportunities or, alternatively, to trigger the buildup of such batches.

This case study also investigated the scalability of the HMES concepts by coordinating manufacturing and transportation activities within networked production. A virtual enterprise was considered, consisting of a network of heat treatment factories. New companies can dynamically join or leave the network and new processes and equipment are introduced as needed. Now the intelligent products have to route their corresponding parts at two levels: the network level and the factory level. At the network level, the intelligent product searches for transportation services between the different factories and heat treatment services (offered by aggregated intelligent resources, offering all services of the resources at a factory). As such, a virtual enterprise is a semi-open system, lacking a single command and control center, the operations have to be organized without the disclosure of sensitive information to other members of the network. Also, a mechanism is required to deal with trust and reputation issues.

The MABE project presented two challenges. First, the effective coordination and control of the highly automated heat treatment facilities critically depends on the ability to exploit the particularities of the equipment and the process plans. It is important that the (expensive) furnaces are (almost) fully loaded with parts. Such loads have to consist of compatible parts, which will be subject to a single/joint temperature trajectory and environmental conditions (e.g., carbon level).

In the existing practice, a high equipment utilization is achieved at the expense of long lead times and high inventories, which allows to find part combinations to fully load the furnaces. The case study had to test and demonstrate the ability to account for the particular characteristics of heat treatment, and it had to achieve this high equipment utilization at lower lead times and/or inventory levels. This amounted to a further validation of the PROSA+DMAS solution.

Second, the main research challenge was to upscale PROSA+DMAS to networked production. It offered the opportunity to construct furnace loads of compatible parts from the entire network. Transportation itself was not addressed by a PROSA+DMAS design, but a simplified model was used.

The actual industrial case was simpler than the research ambition: all networked heat treatment facilities had a single owner (no room for cheating). The research project itself aimed at networked production in which the networked production facilities have different owners, each aiming to optimize their own profit margins, turnover, etc. However, the network is only semi-open. Network members enjoy a trust relationship, which typically is the result of long-term cooperation. From a holonic execution management perspective, the network members are in a highly repetitive game. Therefore, it is not necessary to prevent abuse on beforehand. It suffices to monitor behavior and account for it during subsequent interactions. Abuse by network members is allowed/possible, but it will be noticed.

To this end, the research team elaborated a mechanism and framework to deal with trust and reputation. This enables to assess the information from members in the network. When an order holon (intelligent product) announces to a resource holon (network node) its intention to arrive at 0900 h, this mechanism translates this information in a most likely arrival time (0910 h), the uncertainty on this time (e.g., the 25% and 75% percentiles: 0855 and 1000 h) and uncertainty whether the product will actually visit and use the services of the network node (e.g., 99%). In other words, the design operates much like humans judging how their coworkers behave (e.g., when attending meetings) so they may account for it.

In addition, the enhanced design addressed information disclosure issues by having, for instance, the order holon that is managing network-wide execution trigger the creation of node-internal order holons to manage its node-internal production. The network-wide holon and the node-internal holons exchange information on a need-to-know basis.

Relative to the above-discussed developments, MABE made the following contributions:

- PROSA+DMAS are able to manage networked production facilities, providing manufacturing execution for virtual manufacturing configurations. Moreover, configuring and reconfiguring these virtual manufacturing systems in a supply network was business as usual to the holonic MES.

- It proved possible to enhance the single-factory design to cope with information disclosure requirements and issues, which are encountered in networked production as a result of the presence of multiple self-interested owners. This did not require a fundamental extension as it suffices to apply the pre-existing research results to mimic how humans would handle this (cf. the above discussion of network-wide vs. the node-internal order holons).
- It proved possible to enhance the existing results with a capability to manage trust and reputation, which delivers an artificial/automated social control in the highly repetitive game that characterizes a virtual enterprise. Note that these developments are useful in single-factory cases as well. For instance, it helps to cope with processes that are inherently exhibiting significant and unpredictable variations (e.g., cooking times of vegetables).
- PROSA+DMAS was applied and assessed to yet another manufacturing system class, which again was outside the applicability range of a typical commercial MES. The relevant properties of heat and surface treatment equipment, for an MES, were exotic. Rudimentary data model approaches are unlikely to cope. In contrast, the PROSA+DMAS emulation approach is unlikely to hit the proverbial wall; the required effort nevertheless still depends on the challenge.
- In a short follow-up project, HFID, aimed at technology transfer, emulation of the heat treatment facilities, and its production was generated from multimodels. This generation was done by a self-developed software tool.

Relative to 2015's state of the art, the MABE results have the following limitations:

- In spite of the automated generation of emulation code from multimodels, the effort and time required for a specific case remained (too) high for technology transfer, given that this holonic MES technology is not (yet) established in the market/industry. For our own research purposes and/or for an established MES technology, this required effort and time are workable.
- Technology transfer (still) requires IT skill and service levels that are lacking in many factories. For example, some information resides in spreadsheets for personal use only.

Relative to 2015's state of the art, MPA utilizes the following ICT:

- Software development in Java technology on Windows.
- Emulation/simulation in Java, featuring hybrid emulation.
- Multimodel based generation of Java emulation code.

AgCo2 (Agents for Coordination and Control, 2001–2005)/ ACDPS (Autonomic Computing for Decentralized Production Systems, 2006–2010) (Saint Germain, 2010; Verstraete, 2009; Hadeli, 2006; Holvoet et al., 2009; Parunak et al., 2008; Valckenaers et al., 2009; Valckenaers et al., 2011)

In parallel with the above industry-guided research, fundamental research activities elaborated the PROSA+DMAS research results further. These activities were made possible by extensive financial support from KU Leuven, through their ambitious and very competitive Concerted Research Actions (GOA) program.

In view of the above-discussed developments, the research team made the following contributions:

- The DMAS or delegate multiagent system. The DMAS concept generalizes the nature-inspired stigmergic coordination originating from the MASCADA project and the photographic foil case. Importantly, the terminology for defining and describing this DMAS concept was elaborated, improving communication. In particular, it is important to note the distinction with ant colony optimization, which is a one-shot problem solving/optimization technique (see Section "Ant Colonies and Stigmergy" in Chapter 8).
- The trust and reputation framework, triggered by the MABE objectives, was finalized within these fundamental research activities.
- A generally applicable cooperation scheme between optimizing/scheduling staff holons and the PROSA+DMAS holonic MES has been established. This includes mechanisms that determine whether to follow the staff holon's advice (cf. Section "Cooperation of HMES with Planning Systems" in Chapter 5); note that deviating from the advice at one point in time does not preclude reconnecting to the advice at a later point in time.
- A first contribution toward nervousness control. For the DMAS-generated predictions to be useful, order holons have to resist changing their intentions but have to remain capable of adapting their intentions when justified. The research established three basic mechanisms for individual order holons to manage their nervousness (cf. Section "Socially Acceptable Behaviors for Delegate MAS" in Chapter 5).
- Systematically distinguishing of reality-mirroring software components from decision-making components by the introduction of *intelligent beings* versus *intelligent agents* (cf. Section "The DMAS Architectural Pattern," Chapter 6).

- The ACDPS project translated the concept of Autonomic Computing (Sterrit et al., 2005) originally developed by IBM for automatic software maintenance, to manufacturing and other complex adaptive systems. Autonomic systems manage themselves: they are self-learning, self-optimizing, self-repairing, etc. The project aimed at making complex decentralized production systems autonomic by applying self-X principles. Autonomic behavior can be compared with homeostasis in living systems. Graceful degradation is a consequence of autonomic behavior. A holonic manufacturing system in which a machine breaks down behaves as an autonomic system because it stays active, although in a suboptimal way.
  Future research still may/must address the following (nonexhaustive):
- Nervousness handling in small teams. For instance, order holons may swap allocations and/or reservations. This allows for significant adaptation of their intentions in combination with only minor changes/disturbances for the remainder of the system.
- Generating predictions that employ probabilities and probability distributions. The current developments only generate the expected value. In contrast to the trust and reputation framework, this would be explicitly coded (the trust and reputation framework uses track records to generate expected values with uncertainty information).

Relative to 2015's state of the art, AgCo2 and ACDPS utilize the following ICT:

- Software development in Java technology on Windows.
- Emulation/simulation in Java, featuring hybrid emulation.
- Multimodel-based generation of Java emulation code.

C4AM (Control for Additive Manufacturing, 2010–2011)

C4AM is a technology transfer project. It applied the developments discussed above to an industrial facility for additive manufacturing (a.k.a. 3D printing). This facility has its own MES; commercially available MES offerings cannot cope with the complexities of additive manufacturing. The research team realized a PROSA+DMAS implementation that cooperates with this MES and, specifically, adds the generation of short-term predictions (which was baptized *production radar*). The project successfully demonstrated this production radar. Details are subject to a nondisclosure agreement.

A specific challenge for this project is the requirement for human intervention regarding the composition of a *build*. A build is a composition of products that will be produced together within the work volume of an additive manufacturing machine. Importantly, a significant fraction of the production time remains unchanged regardless of the number of products in a build. Therefore, 3D nesting software is used to optimize productivity when composing these builds. Human supervision and guidance allows significant improvement over a fully automated generation of builds.

As a consequence, the team developed an estimator to compute which product parts will fit in a single build. If the actual build, human-supervised/improved, turns out to be different, the PROSA+DMAS system considers this a disturbance, which is business as usual for this technology. Fortunately, important (high-value) subclasses of products rarely experience this disturbance.

This project experienced a major transition in software technology. Its development started while the research team employed Java on Windows. Because of the unavailability of key personnel at the industrial facility, the final demonstration of the production radar was delayed for several months. In the meantime, the key researcher participated in the MODUM project (cf. below), which was using Erlang/OTP technology. The latter is more robust and scalable by a significant margin. The remaining task for the final demonstration was to connect the Java version to the C# version of the MES.

The researcher decided to sacrifice one day to attempt connecting Erlang/OTP to the MES implemented in C# and "dot-net" technology; it worked within one day. In the following couple of weeks, the Java implementation was reimplemented in Erlang/OTP (note that MODUM already had developed generic software for DMAS in ARTI systems). This swift transition to Erlang/OTP constituted empirical evidence that this high-end software technology is well suited for PROSA+D-MAS implementation. In particular, it reduces the time and effort needed to move from TRL4[2] or TRL5 to TRL7 (and even TRL8) considerably.

The manufacturing cases section concludes with this successful confrontation/marriage of the research results with industrial practice. Next, the research and developments outside manufacturing are discussed.

NONMANUFACTURING CASE STUDIES (Van Belle, 2013; Van Belle et al., 2011a; Van Belle et al., 2011b; Van Belle et al., 2013; Van Belle et al., 2009)

During the final phase of the MASCADA project, an EU-sponsored feasibility study, called MAGECC explored the generic applicability of the MASCADA manufacturing control technology to domains

[2]Technology readiness level.

other than manufacturing. This study delivered a basis for the nonmanufacturing applications that are discussed below; recall that the introduction of this chapter discusses the generic applicability range in more general terms.

Holonic Logistics Execution Systems (HLES)

Logistic operations are neighboring manufacturing (e.g., MABE already addressed networked manufacturing). It therefore constituted a logical next target. In our research, cross-docking has been studied in some detail. Cross-docking is a frequently used logistics strategy. The idea is to transfer incoming shipments directly to outgoing trailers, with little or no storage in between. This practice can serve different goals: cost reduction, consolidation of shipments, etc.

A terminal dedicated for cross-docking is called a cross dock. It is a building with docking slots for trucks. When trucks arrive, they drive backward into their designated slot from which they will be loaded and unloaded. For instance, trucks may arrive from farmers, loaded with a single type of vegetable, and be unloaded into the cross dock. In turn, empty trucks, also docked, will be loaded with a mix of vegetables as they are ordered by the shops. The load distribution is typically such that the unloading at the shops requires minimal reshuffling of pallets. Often, cross-docking is combined with warehousing and quality control.

A general cross-docking case has already been discussed in Section "Cooperation of HMES with Planning Systems" in Chapter 5, where the cooperation of the HLES with a planning system was outlined. An important issue here is the need for multiresource allocation. Indeed, to perform some logistic operations, multiple resources are required at the same time. For instance, to unload an order from a truck at a cross dock, three resources need to be available simultaneously: the involved truck, a dock door to which the truck is docked, and a forklift truck to perform the unloading operation. In fact, for every operation of a forklift truck, two resources are needed: the forklift truck itself and a forklift driver. The HLES deals with this issue by, first, mirroring it in its digital image of the world of interest and, second, apply a search strategy suited for the prevailing situation (cf. Section "Challenges and Lessons Learned from Applications," Chapter 6, on multiresource allocation and auxiliary operations).

Another logistic application (project ELC2) was the holonic control of a chain conveyor. These chain conveyors are commonly used for internal transport in factories, often in complex interconnected network

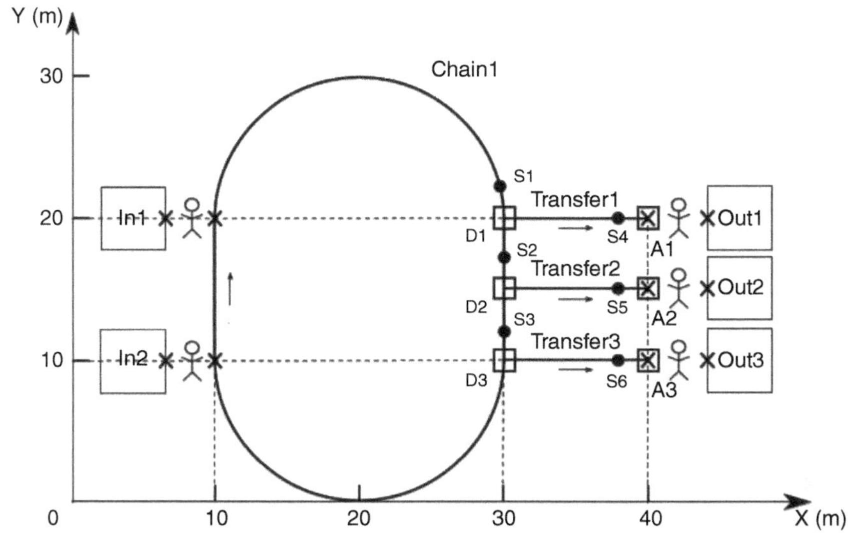

Figure 7.6 ***Layout of a (simple) cross dock using a chain conveyor.***

configurations. In a sample case, a simple application considered a cross-dock distribution center linked by a chain conveyor, as shown in Figure 7.6.

Trucks that arrive at the incoming docks (In1 and In2) are unloaded and the goods are placed on carriers. These carriers are put (manually) on a chain conveyor (Chain1). This chain conveyor has a length of 83 m and has 10 attachment points equally distributed along the length of the chain. Three transfers are connected to the chain (Transfer1, Transfer2, and Transfer3), by means of a diverter. Before each diverter, a sensor reads out the tag of the passing carrier. At the end of each transfer, there is an accumulation stop with a carrier detector in front of it. These accumulation stops hold the carriers until they are removed by a worker and the goods are loaded in a truck at the outgoing docks (Out1, Out2, and Out3). The chain as well as the transfers move at a constant speed of 1 m/s.

Executable domain models were used to forecast, by virtual execution, how the system reacts to a disturbance, in this case, the late arrival of a truck (see also Appendix II on simulation). In a first scenario, five trucks arrive at time 0, one at each dock. The two trucks at the incoming docks, which have to be unloaded, each contain 12 pallets. The three trucks at the outgoing docks are empty and have to be loaded with eight pallets, four pallets from both of the incoming trucks. The carriers can each transport one pallet. The outcome of the simulation shows that the pallets for the different outgoing trucks are mixed up when they are transported by Chain1. The simulation also reveals that the last pallet arrives at its destination after ca. 330 s.

Scenario 2 is identical except for one of the outgoing trucks arriving later than expected. The truck at dock Out3 arrives with a delay of 300 s. The outcome of the simulation shows that the inflow of the pallets is restrained; the pallets for dock Out3 are not released immediately. In the beginning of the simulation, only the pallets for Out1 and Out2 are transported. The last of these pallets reaches its destination already after ca. 210 s. The pallets that have Out3 as destination are released after 250 s. So they arrive at Out3 after 300 s, when the truck has already arrived. The last pallet reaches its destination after 410 s. By controlling the inflow of the carriers, the control system adapts/optimizes the load of the chain conveyors. This desired effect is due to the behavior of the exploring ants when they find out that a resource is not available.

In view of the earlier-discussed developments, the research team made the following contributions:

- The ELC2 project handled chain-based systems in which control actions have long and early commitments (i.e., when placing something on a conveyor chain, the subsequent displacements and occupation of a conveyor slot are fixed until an available removal point is reached).
- The cross-docking developments worked closely with planning systems, to manage the large search spaces.
- The cross-docking addressed multiresource allocation (i.e., loading operations require a truck, a forklift, space, an operator, and the load itself).

Future research may address the following (nonexhaustive):

- Generally applicable solutions for the long commitments as encountered in chain conveyor systems (i.e., a solutions library).
- Generally applicable solutions for multiresource allocation (i.e., a solutions library).
- More robust manners for close cooperation with planning systems. Loose cooperation, having the planner coping with reality, is suitably addressed above (cf. Verstraete 2009). The cross-docking developments selected to be more exposed to planners providing guidance. The project ended before handling all planning cooperation issues could be addressed in full.

Relative to 2015's state of the art, HLESs utilize the following ICT:

- Software development in Java technology on Windows.
- Emulation/simulation in Java, featuring hybrid emulation.
- Multimodel-based generation of Java emulation code.

Robot Fleets (Philips, 2012a; Philips et al., 2012b; Philips et al., 2011)

In Section "The ARTI Reference Architecture," Chapter 6, holonic task execution control of multimobile-robot systems has already been elucidated as an application of the ARTI reference architecture. As could have been observed there, when one wants to benefit from the goodies provided by the PROSA/ARTI/DMAS technology, the environment of the robots has to be included explicitly as a (mostly aggregated) resource holon. For example, Figure 6.4 shows the resource holons for a door opening scenario/task.

The explicit representation and allocation of environment resources facilitates the execution of coordination tasks. Consider the case with an environment consisting of two rooms A and B connected by a narrow corridor. Two wheelchairs, SARA and LAURA, are involved (Figure 6.1); SARA wants to go from A to B, and Laura wants to go simultaneously from B to A. The narrow corridor allows only one robot to pass at a single time; hence, coordination is required.

Without coordination, the mobile robots risk deadlock and live-lock situations (cf. Figure 7.7). Indeed, failure to explicitly manage resource allocation creates the need to enhance this coordination-less system with a deadlock detection and roll-back functionality, which is far from trivial even for a specific system configuration. Without a DMAS mechanism generating predictions concerning the time and duration of resource (narrow corridor) allocations, the mobile robots lack information to balance a longer route against waiting for the resource to become available.

Figure 7.7 (top) shows a trajectory resulting from this lack of information. SARA intends to execute the dashed trajectory. When discovering that the narrow corridor is blocked by LAURA, SARA considers the corridor to be unavailable and recalculates its routing. Next, SARA starts to execute the alternative route circumventing the blocked corridor (if available). When the corridor becomes available again, SARA reverts to its original solution. There obviously is margin for improvement (if only for the wheelchair user to keep his or her confidence in the technology). Centralized planning is one option but not without its known drawbacks (scalability and maintenance issues).

Using PROSA/ARTI/DMAS, the wheelchairs avoid deviations from the shortest path except for collision avoidance (Figure 7.7, bottom). Further simulations reveal that the traveled distances are significantly lower than in the situation without coordination via the environment.

Figure 7.8 shows a slightly more complicated situation. Two wheelchairs, SARA (S) and INGA (G), are to move in opposite directions between two labs

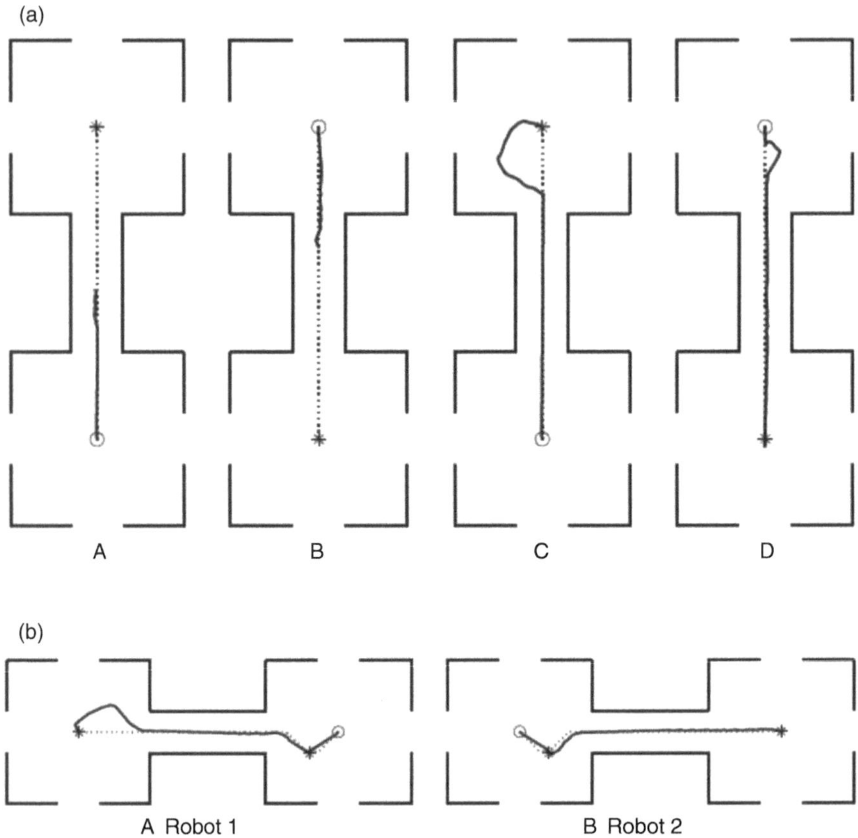

Figure 7.7 (a) Simulation runs of two robots without coordination. Planned (dotted line) and executed (solid line) trajectories from start (circle) to goal (star) are depicted of robot 1 and 2 of a failed (A–B) and a successful (C–D) run. The deviations in the successful run are caused by one robot avoiding the other. (b) Executed trajectories of two robots with coordination of a shared corridor. The robots first move to a transit zone at the side of the corridor's entrance and wait for an available slot on the corridor resource. Whenever a robot has successfully allocated the corridor, it can move through it. The original path, indicated by a dashed line, and the executed path, indicated by a solid line, almost completely overlap inside the corridor.

in the authors' laboratory. There is a mutually exclusive region where coordination is required. Figure 7.8 shows the 3D plot of the trajectories. The solid lines are the executed trajectories with respect to time, and the dashed lines indicate the corresponding trajectory projected onto the environment. The time dimension shows that INGA, starting at the left, waits until SARA, at the right, passes through the area where their trajectories overlap. The fact that the solid trajectories do not intersect indicates no collisions occurred.

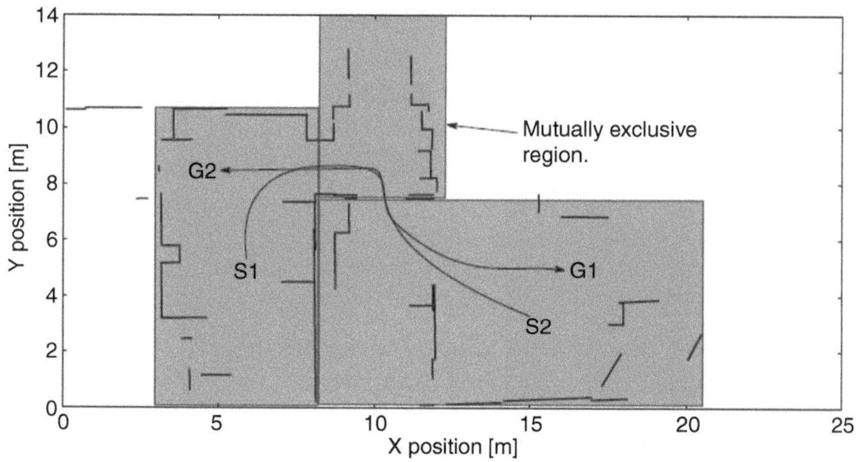

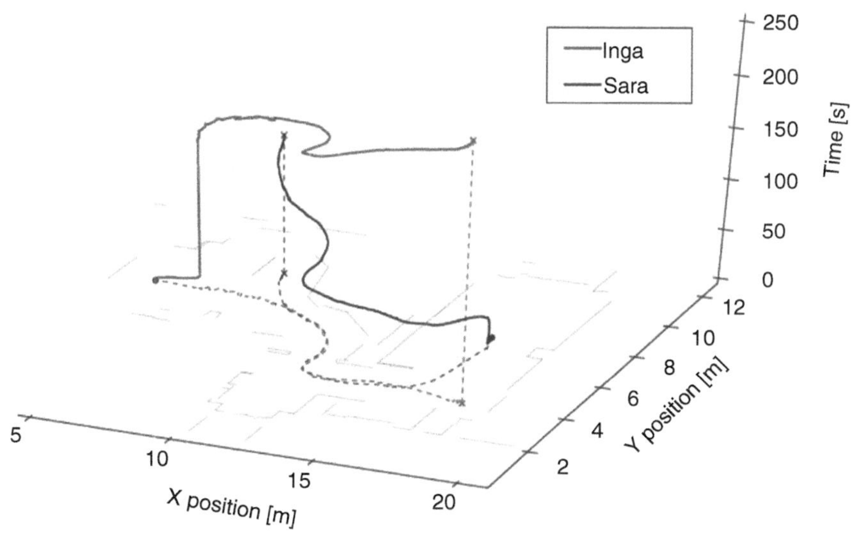

Figure 7.8 ***3D plot of the trajectories followed by both robotic wheelchairs.*** The solid lines are the executed trajectories with respect to time, and the dashed lines indicate the corresponding trajectory projected onto the environment. The time dimension shows how the robot starting at the left (INGA) waits until the robot at the right (SARA) passes through the area where their trajectories overlap. The fact that the solid trajectories do not intersect indicates that no collisions occurred.

In view of the earlier-discussed developments, the research team made the following contributions:

- The research addressed 2D surfaces (possibly on multiple levels) as resources.
- The research applied DMAS to improve the coordination.

- The DMAS prediction assist in preventing deadlock without negatively affecting performance.

  Future research may address the following (nonexhaustive):
- Resource allocation for internal/embedded resources in nonhierarchical fashion (e.g., use of room sensors by a mobile robot, using sensors from another mobile robot).

Open Air Engineering (Ali, 2010; Ali et al., 2013; Ali, EP2531014)

Open air engineering processes, such as open-pit mining, road construction, and farming are mostly carried out with high-tech mobile equipment. This equipment includes self-propelled work vehicles such as excavators, dump trucks, asphalt layers, road graders, combine harvesters, etc. that are specifically designed to carry out these processes (Figure 7.9). Open air engineering processes are capital intensive, and the operating costs of the work vehicles account for a major proportion of the total process cost.

Over the last few years, substantial advancement in the technological development of the work vehicles can be observed. Besides mechanical and mechatronic improvements, an increasing interest is directed toward optimizing the productivity of the work vehicles through proper planning and execution of their operations.

On a generic level, most open air engineering applications share the following:

- The mobile equipment interacts with a surface.
  - These are nonflat 2D surfaces shaped in 3D space.
  - Surface shape and/or properties are modified by the equipment.
- Material is either removed or deposited.
  - Harvester and mining equipment remove.
  - Asphalt layers deposit.
- Some local storage may be available.
  - Asphalt-laying equipment have an on-board buffer.
  - Combine harvesters have an on-board reservoir.
- Material needs to be supplied/shipped from/to a production/processing site.
  - Asphalt is delivered by trucks from producing units.
  - Ore, corn is shipped by trucks to depots and processing units.

  The holonic execution systems need to coordinate these operations:
- The mobile equipment needs to be dispatched to (one of) the sites.

Figure 7.9 ***Open air engineering processes.*** *(Top photograph "Bingham Canyon Mine, west face detail, Utah" by Greg Goebel from Loveland CO, USA - Yibcm_3bUploaded by PDTillman. Licensed under CC BY-SA 2.0 via Wikimedia Commons - https://commons.wikimedia.org/wiki/File:Bingham_Canyon_Mine,_west_face_detail,_Utah.jpg#/media/File:Bingham_Canyon_Mine,_west_face_detail,_Utah.jpg. Middle photograph by Cyron Ray Macey [CC BY 2.0 (http://creativecommons.org/licenses/by/2.0)], via Wikimedia Commons. Bottom photograph "Fertiger-ABG-5820" von Inkulpat aus der deutschsprachigen Wikipedia. Lizenziert unter CC BY-SA 3.0 über Wikimedia Commons - https://commons.wikimedia.org/wiki/File:Fertiger-ABG-5820.jpg#/media/File:Fertiger-ABG-5820.jpg).*

- The "trajectory" of the mobile equipment needs to be determined.
- The transport (trucks) need to be assigned for moving the material from/ to the mobile equipment to/from the processing/production units.

Since the operations of this mobile equipment are expensive, profitability depends directly on how effectively this equipment is utilized. Loss of production capacity is to be avoided and minimized by proper coordination. Idling of the bottleneck resource(s) is the key concern. Here, the coordination faces significant levels of uncertainty and variations in working conditions, possibly shifting the bottleneck from the mobile equipment to the transporters or vice versa.

The yield of the surface processing by the mobile equipment varies and is subject to uncertainty in function of:

- Technical settings of the equipment, representing a trade-off among
  - Getting a lot of work done (e.g., tons of ore or kilos of maize per hour).
  - Energy consumption and/or wear.
  - Risk of damaged equipment.
  - Doing the job well (e.g., leaving very little maize on the field).
- Surface properties:
  - Density of ore.
  - Crop density.
  - Crop type (maize versus beans versus wheat).
  - Shape (slope, corners, etc.).
  - Accessibility (e.g., transporters may not be able to drive across the surface).
- Trajectory executed
  - Parts of the surface may require maneuvering, slowing down the equipment or requiring the equipment to move without processing.
  - Parts of the surface may cause a rough ride, have steep slope.
  - Density variations depending on the trajectory.
  - Accessibility for the transporters of various types.

Coordination needs to adapt to these varying and uncertain working conditions. But, the coordination also impacts on these working conditions. The "trajectory executed" affects how transporters may cross the surface to service the mobile equipment (e.g., trucks cannot drive across a 1 m steep slope created by the excavator, tractors must not drive over still-to-be-harvested crops). Using information about the surface, the trajectory selection affects the expected yield (e.g., crop density will vary across fields where historical data or aerial photography may allow for good estimates).

Overall, coordination – holonic execution systems – can make a difference. When the truck is delayed in traffic, the mobile equipment shall be operated in an energy-saving mode, minimizing losses on the surface, tackling tricky parts of the surface, or perhaps performing small maintenance tasks. When the site is near the processing unit or depot, the mobile equipment utilization is a priority and the truck may have to wait when an optimized trajectory keeps the mobile equipment out of reach until its reservoir is full/empty.

The research investigated the ability of a holonic execution system to generate detailed plans for the cooperating vehicles and to maintain these plans for the changing conditions over time. In the implementation of such on-line planning systems, modeling the planning environment precedes planning and control stages. To model the system environment, the entities in the open-air engineering environment are structured along the ARTI reference architecture.

The novelty in addressing this application domain was twofold. Chronologically, it was among the first to require multiresource allocation and, especially, multiresource allocation for which a leader–follower approach was ill-suited. In particular, the assignment of transporters to activities on mobile equipment prompted the choice for a resource pool holon, managing a collection of very similar resources. Because this research preceded to adoption of Erlang/OTP, the implementation of this approach did not survive the project.

Second, this application domain involved modeling surfaces as resources where activities modify the surface properties and where surface properties determine resource capabilities. For instance, a corn field surface will have initial location-dependent properties based on historical information and measurements. When a combine starts to harvest, properties of the processed parts of the surface need updating (e.g., indicating that tractors may cross). Likewise, measurements by the harvesting equipment can be used to improve the estimates for the unprocessed parts of the surface. DMAS mechanisms explore the processing of these surface parts and, by propagating intention, predict the surface properties in function of time. For example, the exploring ants for a tractor may see which paths to the harvesting equipment will be available when it arrives as well as the estimated position of the mobile equipment.

Overall, coordinating open-air engineering processes involves resource allocation and trajectory determination. Using the DMAS mechanism, this coordination aims to optimize one or more performance objectives, for example, bottleneck utilization, energy consumption, etc. In an open-air engineering process, the work vehicles perform operations at geographically distributed locations (mine site, storage depot, grain fields, etc.). Because of the open and distributed nature of open-air engineering processes, disturbances and variations are highly prevalent in their operating environments.

In practice, plans are generated before the process starts, based on approximate resource performance and predicted operating conditions. Although these plans provide a good starting reference for execution, they are unable to provide the necessary visibility for continued execution of the processes, which are subject to uncertainty and variations. For effective execution, gaining visibility at run time, hence, is imperative. With more run time information, it becomes easier to identify sources of problems or opportunities and take effective decisions. This is what holonic execution systems are designed to provide.

In view of the earlier-discussed developments, the research team made the following contributions:

- The research addressed 2D surfaces (but not necessarily flat) as resources.
- The research addressed 2D surface changing state/properties (e.g., fields get harvested, open air mines get excavated).
- Multiresource allocation was addressed where needed (e.g., combine harvester and tractor executing a rendezvous to transfer wheat or corn). Future research may address the development of generally applicable solutions for multiresource allocation (i.e., a solutions library).

MODUM – Models for Optimizing Dynamic Urban Mobility (Philips et al., 2013; Yperman, 2007)

MODUM was an FP7 Project on intelligent traffic and transportation systems (ITTS). The holonic architecture ARTI and the architectural pattern DMAS were applied to traffic coordination. This led to a *Traffic Radar*, using ARTI to model the traffic infrastructure and several DMASs to forecast traffic flows.

In this holonic architecture, the traffic infrastructure, that is, the road network, is represented by two distinct resource type holons: link holons and node holons. Furthermore vehicle holons, represented by activity instance holons, correspond to trips in the traffic network, originating from traffic users.

A link holon is involved in three distinct actions. First, dynamic map information, received by, for example, a traffic monitoring system, is used to update its state. Second, based on this dynamic map information, for example, indicating traffic density, propagation of flows and cumulative functions are calculated. Third, back-propagation of queues, for example, due to bottlenecks, is performed. Here, D4U is combined with research results from the ITS research community (Yperman, 2007).

Link holons provide D-MAS services to other holons. The *Execute-Scenario* service can be used to perform a what-if scenario on the holon returning travel time on the respective link given an arrival time and estimated traffic density. The *ProclaimScenario* service enables holons to also indicate their intention to arrive on the link. Finally, the *PropagateScenario* service supports adapting upstream or downstream flow on the link.

A node holon is involved in detecting capacity constraints. If flow inconsistency between upstream and downstream links in a particular node occurs, for example, due to an accident bottleneck, the flow has to be adapted. To propagate traffic flow constraints both upstream and downstream, the node holon creates flow ants. This propagation models spill back over links and nodes. The other services offered by node holons are identical to those of link holons.

Vehicle holons represent users driving through the traffic network and, therefore, also virtually move through the link and node holons. They send out exploring and intention ants. These ants drop pheromones on the traffic infrastructure holons to respectively search and proclaim the route of the relevant vehicle holon from its origin to destination.

In view of the earlier-discussed developments, the research team made the following contributions:

- A successful translation of the manufacturing solution to mobility applications.
- The integration of "dynamic network loading models" or DNL models – from ITTS research – into the solution. This delivered a best-of-both-worlds design. ARTI with DMAS is able to propagate user intentions to generate predicted travel routings and road segment loadings. However, a naïve DMAS implementation is ill suited at computing the backpropagation of congestion. Here, the DNL models excel and are highly efficient. Conversely, traffic models are ill equipped regarding forward-propagation (as they have no concept to include detailed user intentions) whereas DMAS excel here.
- Note that when the solution is deployed such that congestions rarely occur in reality (i.e., when they are predicted, user intentions adapt either voluntarily or through regulation when and where the congestion prediction persists), the demands on the DNL models are low; they only are used to stay out of congested states whereas there is no need to

model the precise behavior within such congested states (i.e., no higher-order modeling is needed, dependence on initial/boundary conditions is low, etc.). The merger of results from two domains gives a superior result as both are used where they function well.

Future research may address the following (nonexhaustive):

- The application of ARTI with DMAS to the public domain revealed the need for accompanying innovations in the social domain. Indeed, the developed solutions yield the most benefits when there is a high level of participation in combination with a – beneficial level of – social control. Specifically, the prediction generation through delegate MAS needs high participation (the nonparticipating parties are considered and treated as disturbances). And the social control prevents good deeds from being punished; travelers accommodating others get rewarded for their contribution. In fact, they can refrain from accommodating until they have received guarantees for their reward.
- The project ended before it was able to address multimodal transport in full. However, the nature of a D4U solution makes it plausible that it only is a matter of modeling the world of interest in sufficient detail. Note however that this open issue shares the multiresource allocation challenges mentioned earlier already more than once.

Relative to 2015's state of the art, HLES utilize the following ICT:

- This project made the transition to Erlang/OTP. It was a drastic improvement.
- Note that much of the unfinished items in the other cases, discussed earlier, can be allotted to the difficulty/impracticality of addressing them in the Java implementations.

Railway Operations (De Swert et al., 2006)

On a much smaller scale, the team looked at railway systems. Again, the issue of handling long commitments emerged (Cf. ELC2 project). In railway operations, with trains being unable to overtake, the challenge of anticipating the impact of decisions needs addressing, for instance by exploring and/or intention ants making roundtrips, collecting information on the way out and building a solution/journey on the way back.

Smart Grid (Rutten and Valckenaers, 2013)

Also on smaller scale, smart grid applications were investigated. The research looked in active demand at the smallest granularity possible, acknowledging that aggregating afterward is far easier than the reverse (in IT systems). It foremost was an exercise in covering continuous-time domains:

- Whereas products in a factory and travelers on the road execute a trajectory visiting resources (e.g., processing equipment and road segments respectively), electricity consumers and producers have a time-continuous profile for their consumption or production, respectively. Consequently, intentions will be profiles in function of time of power, energy, etc. Note that the smart grid community introduced the notion of prosumers, which may both produce or consume.
- Electricity is a single commodity. Here we point out that electrons injected at one point in the grid, sold by producer P, do not have to be transported and distributed to consumer C, who bought this from P. Electrons are exchangeable, which is not true for the above applications.

This offered an interesting challenge for the application of the PROSA+DMAS concepts. It proved possible to design an exploring DMAS as well as an intention-propagating DMAS, not only for energy or power consumption but equally for the distribution of this power across phases, for voltages, for reactive power, etc.

Interacting with the smart grid community revealed that the above remains foremost an exercise in applying D4U to an application domain with radically different properties, showing that it is possible to effectively use the D4U concepts and approach in such domains. What remains to be investigated and described is where and how D4U is best applied and used within a smart grid. Or in other words, where are the conditions for a successful application fulfilled in a socioeconomic sense? And where and how can D4U be introduced within the comfort zones of the grid community?

First of all, note that the economic gains from applying D4U in a smart grid are limited in comparison to, for example, the potential benefits in manufacturing or health care. Energy still is relatively and surprisingly cheap (as is transport in logistics). D4U gains come from exploiting the demand-response margins, not from the energy generation and consumption per se. This represents only a fraction of the economic value in the electricity grid. Larger gains have to come from "not having to invest in building new infrastructure" (e.g., power lines). Moreover, consumers may avoid discomfort by paying more (unlike traffic where this only brings better in-traffic-jam entertainment) as long as blackouts are avoided.

Within the grid and from a D4U perspective, there roughly are three distinctive areas:

- Transmission (TSO[3] domain)
- Distribution (DSO[4] domain)
- Home domain.

Their suitability for D4U application differs considerably.

Transmission Network Operations

In the TSO domain, power is transported along a mesh of power lines at high voltages. Only large installations (generators, industrial installations) are connected to this highest level of the grid. From a D4U perspective, this is a small system representing large investments serving a massive number of customers. Hence, conventional manners of operating will most likely remain effective and efficient. It is economically feasible to have teams of (well-paid) human experts manage the transmission system with the assistance of ICT-based planning and/or control systems in shifts covering 24 h every day of the year. It is economically feasible to install (very) expensive sensors and actuators.

Moreover, the meshed nature of transmission networks requires a D4U design that is similar/analogous to the MODUM design discussed above. To model and compute the state of a power transmission network, the applicable laws (i.e., Kirchhoff's laws) need to be accounted for. Existing systems in the TSO domain are able to provide suitable services, which compute the states and available capacities of the network elements over (near-future) time (i.e., when provided with the boundary condition values).

These TSO services must then cooperate with an ARTI and DMAS subsystem, which is using the (predicted) available capacities to explore for solutions and reserve capacity for the prosumers in the smart grid. The intention propagation provides inputs for the TSO services, allowing them to compute the network state for future (predicted) boundary conditions (i.e., what consumers, producers, and prosumers intend to do, propagated by DMASs through their connection by their distribution system).

This may result in predicted improper states (e.g., overloading a power line or voltage transformer some 3 h into the future). The refresh mechanism of the D-MASs will observe this and have the activity holons adapt.

[3]Transmission System Operator: organization responsible for the high-voltage long distance power grid.

[4]Distribution System Operator: organization responsible for a medium- to low-voltage part of the grid, sitting between the end user (homes) and (a connection to) the high-voltage transmission system.

Here, timely convergence has to be ensured; note that this can be done by a virtual safety net that simply will be picking suitable victims (when the self-organizing mechanisms remain ineffective).

However, the economic and technical benefits of applying the above cooperation are doubtful. Currently, considering the transmission systems to be copper plates connecting the distribution systems is likely to be sufficiently accurate for coordination with DMAS on a grid scale. The errors would be manageable disturbances. More in general, "balancing production and consumption" first and handling transport second (as a control problem) appears to be a feasible approach in the grid at this level (recall that power/electricity is single commodity).

Smart Homes

On the other extreme, the home level struggles with the (meagre) amount of power (flexibility) on offer and the amount of ICT investment needed for these small amounts of power. However, in the long run, these ICT investments will be shared with other concerns (home security, comfort, health care), and perhaps even cost savings when electrical appliances use their networking capabilities to have their control panels solely on smartphones, tablets, and computers (simplifying the device itself).

Here, the system is huge, which implies that conventional approaches will not work. The question is how to make them cooperate with the conventional system (at the TSO and DSO levels). For instance, instead of developing a full-blown D4U execution system, D4U insights can be used to make flexibility available to the more conventional solutions. This offers two kinds of benefits.

First, it postpones as long as possible the conversion of a representation of the actually available flexibility into a format used by the prevailing conventional problem-solving mechanism. This allows to change, improve, or replace this problem-solving mechanism with little effort (i.e., the early mechanisms do not become a legacy issue). It also makes visible/known what the potential improvements for such mechanisms are when the information reflects the actual flexibility (and not an information-losing projection on the format needed by the current problem-solving mechanism).

Second, a D4U representation of actual flexibility will mirror how this connects to the nongrid side. It will provide a suitable starting point to coordinate, across boundaries, between the electricity grid and, for instance, the system that is using the electricity to cool, heat, cook, clean, charge, etc. Moreover, making flexibility explicitly available has the advantage that it can be done beforehand (e.g., soft real time over the internet), whereas actually using this flexibility can happen very fast. Indeed, it can be used as reserve power to manage an imbalance between supply and demand within seconds.

Distribution System Operations

The DSO domain is facing serious challenges. Individually, they may be modest-size systems from a D4U perspective but they represent smaller investments and serve fewer customers than in the TSO domain. Cost-effectiveness is an issue[5] when upgrading the current systems to address the future challenges posed by renewable energy and prosumers.

It is undecided whether the DSO domain will have the option to address its upcoming challenges within its own domain. It may need to drastically improve its sensing and measuring capabilities as well as it actuation and steering capabilities. But inside their own systems, the required investments (number of sensors and actuators multiplied by their unit costs) appears to be prohibitively high. Smart solutions, implying cooperation with nonDSO entities, appear to be superior concerning cost-benefit.

Here, research may look into rendering a large number of low-quality sensors, actuators into a high-quality aggregated sensor-actuator. The single commodity property offers possibilities. However, these low-quality (consumer-grade) sensors and actuators will reside in the home domain. Possibly, a low number of high-quality industrial-grade sensors may serve to calibrate and complement the low-quality ones. Note that these consumer-grade sensors in networked electrical compliance may become very cheap if this is planned and standardized appropriately (i.e., integrated on the device control IC).

In addition, some kind of D4U execution system may be needed to turn a large number of small actuators into an equivalent of a high-quality actuator. Here, single commodity implies that this actuation may reprofile and must not be responsible to the actual full profile. Here, the DSO network is not meshed at any given instance. (It can be reconfigured and has the potential for meshing, but this would only complicate its control without offering an incentive to do so.) This provides a natural environment for DMASs.

Discussion

Overall, it still are early days for D4U in smart grids. Conditions for smart solutions will be improving for the foreseeable future. It will be possible to share costs with other application domains, and the growing importance of renewable energy, electrical cars, heat pump, CHP installations, etc. will render

[5]Looking at one's electricity bill, distribution accounts for a most significant percentage.

smart solutions more viable. The speed and the magnitude of this shifting will determine what kind of solutions will be needed. For instance, current and forthcoming practice relies on the fact that electricity is a single commodity and the aggregated demand of a massive number of users averages into a fairly predictable profile (in function of date, time, and weather). When the intelligence in the grid and its users increases, these predictions may no longer remain valid, and users may need to communicate their intentions and commitment levels to preserve a usable forecast (as we have today).

Today, the smart grid community is solving its future challenges as they present themselves by adequate solutions that respect the prevailing players as good as possible. Importantly, preventing blackouts and providing balancing power will and can be addressed by shifting household consumption without a lot of intelligence (e.g., increase fixed day–night prices into four to five fixed time periods) and demand response residing at big (industrial) prosumers.

Further increasing renewable power production is likely to be achieved more economically by keeping sufficient conventional installations in the stand-by mode (both spinning reserves and otherwise) than by massive intelligence in households in the immediate future. Home intelligence investments need to be extremely low cost (i.e., integrated in future products rather than installed separately on existing). Or it needs to be paid for by other parties/concerns than shifting electricity consumption. For example, home intelligence for security, comfort, and health care may also be usable for energy management purposes and provide household demand response almost for free.

In conclusion, this research was interesting because of the radically different nature of the application domain. However, the economic reality makes it unlikely that fine-grained demand response will become a reality in the foreseeable future. Electrical energy, electricity, remains relatively inexpensive, and cherry-picking, utilizing the best and least complicated/discomforting opportunities first, is likely to keep the grid up and running. Note, however, that design for the unexpected goes beyond grid intelligence for demand response. Forthcoming research addresses more technical aspects concerning interoperability (cf. Section "Smart Homes, Smart Grids, and Energy Storage").

ONGOING AND FORTHCOMING CASE STUDIES

Current and forthcoming research is expanding the applicability of design for the unexpected into two domains. The first is e-health. The second is smart homes and interoperability; the focus is on storage of energy, both electrical and thermal.

E-health and Integrated Care (Valckenaers and De Mazière, 2014)

Recent ongoing research has elaborated a conceptual HHES[6] design[7] targeting "integrated care and multidisease." Applying ARTI and DMAS, patients and care providers are "extended" and "enhanced" by reality-mirroring computer processes. For instance, there will be, for each patient or care provider, an e-Person/holon comprising

- A resource type holon,
- A resource instance holon,
- An activity type holon,
- An activity instance holon.

These will be composite/aggregated holons, where each holon may be, in turn, divided into intelligent beings (executable software models mirroring reality) and intelligent agents (modeling the decision making).

A patient will be considered an aggregated resource in both the logistic and medical sense. In the logistic sense, a patient must be, for example, present for a medical intervention, must be conscious to answer questions, must be physically able to self-inject insulin, etc. In the medical sense, a patient is considered a composite resource comprising kidneys, a liver, a stomach, a blood circulation system, etc. where, for example, these organs need to have the capacity to tolerate medication and other treatments. Care providers will be considered resources required to execute health care activities. Other resources, reflected in the HHES, correspond to equipment (e.g., a CT scanner) or supplies (or supply channels) for medication and other consumables.

Health care activity instances use exploring and intention ants to coordinate care proactively, even across organizational boundaries. Health care activity types inform instances about the available options (e.g., self-inject insulin or have another person inject the insulin). The activity instances search for and recruit suitable resources to execute one of the options. They do this proactively, discovering problems and opportunities early.

When the patient is injured and cannot self-inject insulin, the instance looks for a person to assist as soon as information about the injury becomes known. When in case of multidisease multiple medications are about to enter the patient's stomach or bloodstream together, the intention-propagating DMASs will inform the affected resource instance holons as soon as the

[6]Holonic Healthcare Execution System.

[7]TRL2.

corresponding activity instances are activated (and send out intention ants). This information can be fed to a triage mechanism (implemented as an intelligent agent) to check for interaction among these medications. This may change the "predicted" result of an activity instance step when executed virtually, causing the activity types to account for interactions.

In health care, the challenge addressed in C4AM returns in full. The solution needs to distinguish – that is, provide separate but paired software components – between best-guess and signed-off implementations (cf. Section "Challenges and Lessons Learned from Applications," Chapter 6). To generate predictions efficiently, fully automated software of low computational complexity must and will be employed. However, when decisions and actions are for real, suitably authorized humans are to be involved (and in control).

Furthermore, privacy and information disclosure become key concerns. MABE already delivered insights and mechanisms but further developments are needed. Importantly, our solution – comprising communicating computing processes – offers interesting opportunities and possibilities. Indeed, it is possible to imitate human-based solutions in which two agents share information on a need-to-know basis.

Moreover, the open nature of e-health applications, in particular our patient-centered and empowering design in integrated care, calls for novel ICT-enabled cooperation. In fact, this domain is highly suited to investigate social innovations leading to new beneficial/desired manners for people to live together. It is a vehicle to discover what is needed to create a warm synergy-enabling mechatronic society.

Overall, e-health and especially integrated care crossing organizational borders are suitable targets for the design for the unexpected. It has the right properties, and it offers interesting challenges to trigger insights and the development of capabilities.

Smart Homes, Smart Grids, and Energy Storage

Recently, D4U research in home automation and particularly in energy storage focuses on devices and systems of devices, addressing in-depth interoperability. Here, interoperability is understood to enable far-reaching integration achieved by establishing the proper connectivity without having to redevelop the components, devices, or systems themselves.

The research activities look at devices as a structured collection of resources, on which activities execute. In-depth interoperability implies that those resources remain accessible, also for unexpected uses. If the native device controls fail to support such unexpected use, interoperability support signifies that its resources can be deallocated from the native controls and allocated to another one. Furthermore, resource instances need an agenda service to enable proactiveness (provided by an intention-propagating DMAS).

Moreover, device services lacking critical mass also have to allow deallocation and replacement when and where desired. This applies most definitely to the programming services and facilities offered and supported by a device or system. The research will describe and communicate that there exists a "valley of death" for programming facilities/languages/tools (offered on devices).

On one side of this valley, the KISS principle is upheld. Only very simple services are available, which can be used without a lot of training or expertise/talent and which pose few debugging challenges. Here, in-depth interoperability typically requires external systems to have access to the sensors, actuators, and state information of the device with sufficient bandwidth and low delay. Indeed, the device services are lacking the expressivity needed for nontrivial applications.

Older industrial automation technologies reside on this side of the valley but they are very close to the edge (as they are programming tools), and the valley is expanding in manners threatening to engulf them. Human talent, capable of developing nontrivial applications, may only be motivated externally (e.g., by high wages or positions in the organization) to dedicate time and effort to these technologies. The survival of these industrial technologies can be ensured by shrinking their responsibilities such that they increasingly comply with the KISS principle (i.e., the technology can be used by most employees knowledgeable about the application without needing much programming skills or expertise).

On the other side of this valley, full-fledged programming languages, tools, and platforms are used. They need critical mass (cf. Chapter 4). In particular, talented computer scientists – or equivalent – must either already master the programming technology or must be self-motivated to learn how to use it. If it is necessary to motivate them and keep them motivated by external means (e.g., by high wages or by a position in the organization that cannot be given to all the members of a team), it is advisable to conclude that the programming technology resides somewhere in this valley of death.

The programming technology of choice in our more recent research, Erlang/OTP, resides on this other side of the valley but sits still close to its edge. Mainstream technologies (C/C++, C#, Java, Python) are situated at a safe distance today but the valley is shifting. It is shifting in directions that favor Erlang/OTP as distributed and massively multithreaded programming rapidly are becoming the norm. Furthermore, this technology offers advantages over mainstream alternatives that are relevant and even decisive. For instance, it significantly reduces the effort to transform a lab prototype into deployed solution.[8] Moreover, the robustness and stability of Erlang technology implies that it is sitting on very firm ground next to the valley, allowing it to survive for more than 25 years.

In contrast, the newer sophisticated technologies from industrial automation are situated inside this valley, which is expanding in manners that make this situation worse for the foreseeable future. The absence of mature open source support is a tell-tale sign for many of these technologies. Indeed, how does industry expect a widespread adoption by teaching institutes when it is behind payment walls or requires time-consuming[9] negotiations to get free access for ICT or industrial automation courses? Likewise, many technologies from the academic communities, including the artificial intelligence or multiagent communities, do not appear to escape from this valley.

In general, it is a decisive and nontrivial matter to decide which programming technologies to embrace. Too conservative is a sure death in the longer run. Too adventurous equals risking death in the shorter run. It inherently is a balancing act (Waldrop, 1993) confirming that life indeed resides in the small region between order (too conservative) and chaos (too adventurous).

Concerning the activities that execute on the devices, which are implemented with these programming tools, D4U principles require to make flexibilities available. In this respect, research needs to discover what the agenda services of devices have to offer. The discussed explorations in the smart grid domain will be continued here. Overall, the smart homes and energy storage research constitutes an instrument to design and describe (mechatronic) device control architectures and systems that are designed for the unexpected.

ABBREVIATIONS

ARENA®	Discrete event simulation and automation software
C4AM	Control for additive manufacturing
CHP	Combined heat and power
DSO	Distribution system operators
FACCS	Flexible assembly cell control system
HHES	Holonic healthcare execution system
IC	Integrated circuit
TRL	Technology readiness level

[8] Going from TRL4/5 to TRL7/8.
[9] Time from ICT lectors and teachers.

REFERENCES

Ali, O., 2010. Operational planning for outdoor engineering processes, Doctoral thesis, KU Leuven.

Ali, O., Valckenaers, P., Van Belle, J., Saint Germain, B., Verstraete, P., Van Oudheusden, D., 2013. Towards online planning for open-air engineering processes. Comput. Ind. 64 (3), 242–251.

Ali, O., Valckenaers, P., Van Belle, J., Saint Germain, B., Verstraete, P. In use adaptation of schedule for multi-vehicle ground processing operations – EP2531014, US20130046525, WO2011095614, European patent register.

De Swert, K., Valckenaers, P., Saint Germain, B., Verstraete, P., Hadeli, K., Van Brussel, H., 2006. Coordination and control for railroad networks inspired by manufacturing. In: Proceedings of the IEEE Workshop on Distributed Intelligent Systems. IEEE Workshop on Distributed Intelligent Systems. Prague, Czech Republic, June 15-16, 2006, pp. 201-206.

Hadeli, K., 2006. Bio-inspired multi-agent manufacturing control systems with social behaviour, Doctoral thesis, KU Leuven.

Holvoet, T., Weyns, D., Valckenaers, P., 2009. Patterns of delegate MAS. Self-Adaptive and Self-Organizing Systems. In: IEEE Computer Society. International Conference on Self-Adaptive and Self-Organizing Systems. San Francisco, CA, USA, September 14-18, 2009, pp. 1-9.

Parunak, H., Brueckner, S., Weyns, D., Holvoet, T., Verstraete, P., Valckenaers, P., 2008. E pluribus unum: Polyagent and delegate MAS architectures. In: Lecture Notes in Computer Science, Vol. 5003, Springer, Berlin, Heidelberg, pp. 36–51.

Peeters, P., Van Brussel, H., Valckenaers, P., Wyns, J., Bongaerts, L., Kollingbaum, M., Heikkilä, T., 2001. Pheromone based emergent shop floor control system for flexible flow shops. Artif. Intell. Eng. 15 (4), 343–352.

Philips, J., 2012. Holonic task execution control of multi-mobile-robot systems, Doctoral thesis, KU Leuven.

Philips, J., Valckenaers, P., Aertbeliën, E., Van Belle, J., Saint Germain, B., Bruyninckx, H., Van Brussel, H., 2011. PROSA and delegate MAS in robotics. Holonic and Multi-Agent Systems for Manufacturing, Vol. 6867, Springer, Berlin, Heidelberg, pp. 195-204.

Philips, J., Valckenaers, P., Bruyninckx, H., Van Brussel, H., 2012. Scalable and robust coordination of multiple mobile robots using PROSA and delegate MAS. International Symposium on Robotics, pp. 527-532.

Philips, J., Saint Germain, B., Van Belle, J., Valckenaers, P., 2013. Traffic radar: a holonic traffic coordination system using PROSA++ and D-MAS. In: Lecture Notes in Computer Science, Subseries Lecture Notes in Artificial Intelligence. Industrial Applications of Holonic and Multi-Agent Systems, Vol. 8062, Springer-Verlag, Berlin, Heidelberg, pp. 163-174.

Rutten, L., Valckenaers, P., 2013. Self-organizing prediction in smart grids through delegate multi-agent systems. In: Bajo Pérez, J., Corchado Rodríguez, J., Fändrich, J., Mathieu, P., Campbell, A., Suarez-Figueroa, M., Ortega, A., Adam, E., Navarro, R., Moreno, M. (Eds.), Advances in Intelligent System and Computing, Vol. 221, Springer International Publishing, PAAMS, Salamanca, June 23-25, 2013.

Saint Germain, B., 2010. Distributed coordination and control for networked production systems, Doctoral thesis, KU Leuven.

Saint Germain, B., Verstraete, P., 2002. Multi-agent fabrieksbesturing in Java, Master thesis KU Leuven, (in Dutch).

Saint Germain, B., Valckenaers, P., Van Brussel, H., Van Belle, J., 2011. Networked manufacturing control: an industrial case. CIRP J. Manuf. Sci. Tech. 4 (3), 324–326.

Saint Germain, B., Valckenaers, P., Van Belle, J., Verstraete, P., Van Brussel, H., 2012. Incorporating trust in networked production systems. J. Intell. Manuf. 23 (6), 2635–2646.

Simon, H.A., 1990. The Sciences of the Artificial. MIT Press, Cambridge, MA.

Sterrit, R., Parashar, M., Tianfield, H., Unland, R., 2005. A concise introduction to autonomic computing. Adv. Eng. Inf. 19, 181–187.

Valckenaers, P., De Mazière, P., 2014. Innovative ICT Systems for Integrated Care. Proceedings Med-e-tel, global telemedicine and ehealth updates: knowledge resources 7, 464–468, ISSN: 1998-5509.

Valckenaers, P., Van Brussel, H., Bongaerts, L., Bonneville, F., 1995. Programming, scheduling, and control of flexible assembly systems. Comput. Ind. 26, 209–218.

Valckenaers, P., Saint Germain, B., Verstraete, P., Van Belle, J., Van Brussel, H., Hadeli, K., 2009. Intelligent products: agere versus essere. Comput. Ind. 60 (3), 217–228.

Valckenaers, P., Van Brussel, H., Bruyninckx, H., Saint Germain, B., Van Belle, J., Philips, J., 2011. Predicting the unexpected. Comput. Ind. 62 (6), 623–637.

Van Belle, 2013. A holonic logistics execution system for cross-docking, Doctoral thesis, KU Leuven.

Van Belle, J., Saint Germain, B., Verstraete, P., Valckenaers, P., Ali, O., Van Brussel, H., Cattrysse, D., 2009. A holonic chain conveyor control system: an application. International Conference on Industrial Applications of Holonic and Multi-Agent Systems, pp. 234-243.

Van Belle, J., Saint Germain, B., Valckenaers, P., Van Brussel, H., Bahtiar, R., Cattrysse, D., 2011. Intelligent products in the supply chain are merging logistic and manufacturing operations. Eighteenth IFAC World Congress, pp. 1596-1601.

Van Belle, J., Valckenaers, P., Saint Germain, B., Bahtiar, R., Cattrysse, D., 2011. Bio-inspired coordination and control in self-organizing logistic execution systems. Ninth IEEE International Conference on Industrial Informatics, pp. 713-718.

Van Belle, J., Saint Germain, B., Philips, J., Valckenaers, P., Cattrysse, D., 2013. Cooperation between a holonic logistics execution system and a vehicle routing scheduling system. In: Eleventh IFAC Workshop on Intelligent Manufacturing Systems, vol. 11(1), pp. 184-189.

Van Belle, J., Philips, J., Ali, O., Saint Germain, B., Van Brussel, H., Valckenaers, P., 2014. A service-oriented approach for holonic manufacturing control and beyond, section 3.1.2. In: Service Orientation in Holonic and Multi-agent Manufacturing and Robotics, Springer, Berlin, 2014.

Van Brussel, H., Wyns, J., Valckenaers, P., Bongaerts, L., Peeters, P., 1998. Reference architecture for holonic manufacturing systems: PROSA. Comput. Ind. 37, 255–274.

Verstraete, P., 2009. Integrating existing scheduling techniques into the holonic manufacturing execution system, Doctoral thesis, KU Leuven.

Waldrop, Mitchell M., 1993. Complexity: The Emerging Science at the Edge of Order and Chaos. Simon and Schuster, New York, 380 pages.

Yperman, I., 2007. The link transmission model for dynamic network loading, PhD thesis, KU Leuven.

Zamfirescu, C., Valckenaers, P., Hadeli, K., Van Brussel, H., Saint Germain, B., 2003. A case study for modular plant control. In: Holonic and Multi-Agent Systems for Manufacturing, Springer, Berlin, pp. 268–279.

CHAPTER EIGHT

Work by Others

Paul Valckenaers*, Hendrik Van Brussel**
*Faculty of Engineering Technology, KU Leuven
**Faculty of Engineering Science, KU Leuven

This chapter discusses work on multiagent (manufacturing) execution systems by others. It is not our intention at all to provide coverage or completeness for the research domain. Moreover, the inclusion or omission of contributions by other research teams does not reflect their importance or impact. The sole goal of our selection is to deepen the readers' understanding of this book. In particular, the section aims to highlight differences and similarities while discussing the trade-offs involved as well as the context, which induced these other teams to make their choices.

PRODUCTION 2000+ (Bussmann et al., 2004; Schild and Bussmann, 2007)

This research is well documented at www.stefan-bussmann.de/en/agents/p2000p.html. Its agent-based control system comprises three types of agents:

- *A work piece agent* for every work piece in the system
- *A machine agent* for every machine tool of the system
- *A switch agent* for every routing element in the transport system

The work piece agents – knowing the state and processing graph of their work piece – look for processing capacity to execute their "operations to be performed next." The allocation of a next machine is carried out by a simple first-price, single-round auction.

The protocol is initiated by a work piece agent. It determines the next operations to be performed as well as a list of all the machines that are configured to perform (part of) these next operations (by looking into a static configuration list). The work piece agent then sends an invitation to bid, which includes a specification of the operations to be performed, to all the machines in the list.

Next, the machine agents send a bid, including (i) the current size of their virtual buffer and (ii) the maximal set of operations they can perform during a single visit by the work piece. The work piece agent then collects

Design for the Unexpected. http://dx.doi.org/10.1016/B978-0-12-803662-4.00008-4

all bids and awards the best bid. Both components of a bid are used where the current size of the machine's virtual buffer has the higher priority (to balance the workload). The awarded machine then includes the work piece in its input buffer.

This allocation protocol (www.stefan-bussmann.de/en/agents/p2palgorithm.html#allocation) is actually more elaborate. Importantly, this protocol was designed for the following operating range, in which good performance will be achieved:

- A flexible transportation system able to move a work piece from any machine to any other machine as required by product variants.
- Flexible machines providing a range of operations to produce any variant of a product type.
- Large-series manufacturing in which a single work piece must be finished at some time, but not in the shortest time possible and not with a relative priority.

Overall, these are conditions in which a local and myopic decision is competitive with planning ahead. Note also that major characteristics of the underlying production system (e.g., flexible transport) are "hardcoded" in the design. And data formats are exchanged that inherently make assumptions about the process plans (product holon) and resource (type) capabilities.

The control system development itself applied the following steps:

1. *Analysis of decision-making.* The team identifies and analyses the control *decisions* that are necessary to operate the manufacturing process. Also, dependencies between the decisions, causing a need for interaction during the execution, are identified and incorporated into a decision model.
2. *Identification of agents.* This step determines the system architecture of the agent-based control system. In particular, it decides about the agents in the control system while focusing on the *decisions* they are responsible for (including the interaction needed for these decisions).
3. *Selection of interaction protocols.* From a library of *existing* (agent-oriented) interaction protocols, a (most) suitable interaction protocol is selected and, as required, adapted to the specific needs of the particular case.

The development approach is based on the observation that, in industry, every manufacturing process requires a different kind of flexibility and functionality; there is no one-fits-it-all agent-based control system. When designing a control application, designing/customizing the agents of the control system must be part of the implementation process. This observation is correct: as a rule, manufacturing systems will have special needs relating to and, when decisions are the concern, impacting on a company's competitiveness.

The adopted approach leaves all participants in their comfort zone. Its objectives include providing methodologies that enable an ordinary[1] (manufacturing) engineer to apply agent technology without being an expert in agent technology. It also does not encourage the IT specialist to capture insightful manufacturing expertise and knowhow in software components or systems; reflection of the world of interest, as in D4U, is not prominently present in the approach or methodology.

The project was extremely successful concerning technical performance improvement over conventional dedicated production lines using hard automation. Nonetheless, the agent-based control system was not adopted beyond the initial manufacturing plant. The company learned which flexibility is most beneficial and, subsequently, introduced more targeted flexibility in more conventional production systems using industrial automation software technologies. This was less expensive and could be done with readily available (human) expertise. Because too many properties of the ultra-flexible manufacturing system in Production 2000+ had been "hardcoded" into the agent-based control system, this system could not adapt fast enough and, thus, was unable to compete against solutions that benefited from insights generated by this project.

Discussion

Production 2000+ advocates the complete opposite of *design for the unexpected*. Its approach and development methodology is based and focusing on the decision making whereas in *design for the unexpected*, reality mirroring in executable models constitutes this primary concern. In D4U, decision-making elements are introduced late and the overall system must never become highly committed to them.

The inability of Production 2000+ to adapt to a different underlying (manufacturing) system played a key role in its demise. However, a D4U solution would have been significantly more complex, if only because of the DMAS implementations that it uses to have its decentralized decision making compete with more conventional planning ahead. Applying the KISS principle did not save the DACS[2] methodology proposed by Bussmann. Agent interaction protocols, which are its main source of reusable software, deliver

[1]From the perspective of the ICT developers in the Production 2000+ team.
[2]The design methodology proposed by Bussmann from his Production 2000+ experience.

far too little of the overall functionality that is required from a control system. More reuse of software and more sharing among multiple users is needed.

D4U and DACS are opposites. But this does not imply that either needs to be wrong. The DACS approach and steps in Production 2000+ are well suited for a final development phase: they come last and implement the features and functionality as requested by the user. It is suited to finish the job. In contrast, design for the unexpected is superior to develop the reusable platforms on top of which a Production 2000+ approach builds a specific working system. Note that, when combining D4U and DACS, this final phase develops pure decision making components, not full-fledged agents.

XPRESS (XPRESS, 2007–2011)

XPRESS – Flexible Production experts for reconfigurable assembly technology – is a so-called Integrated Project funded by the European Commission under FP6-NMP. Its research has been continued in an academic setting and published in Peschl (2014). The goal of XPRESS was to establish a flexible production concept based on the idea of "specialized intelligent process units" (called expertons) for customized production. XPRESS aimed to integrate intelligence and flexibility at the highest level of the production control system as well as at the lowest level of the singular machine. The expertons have been renamed into manufactrons for "legal" reasons (some other party already possessed rights on the experton name).

Developments Driven by End User Requirements

In contrast to D4U, which started from a theoretical understanding of the root causes for failure during integration, XPRESS starts from identifying and consolidating industrial requirements. While starting from end user requirements is not a panacea for success, these requirements reflect the desiderata of the industrial automation and manufacturing communities as well as their (lack of) insight. They constitute a valuable research result as such.

A common risk, when using user requirements to drive developments, are overly tight connections to these requirements. For instance, consider the requirement for deterministic behavior. This normally is achieved by constraining what the system elements may do. However, such constraining can be introduced late. A user requirement for deterministic behavior implies that early developments may not prevent such constraining but does not mandate that such constraining needs to be introduced at all levels and at all times.

Wishful thinking – the Christmas list syndrome – is another issue, with user requirements driving development too tightly. Technology hyping is causing considerable damage whenever user expectations prove illusionary. For instance, looking again at the requirement for deterministic behavior, a highly constraining design may – deterministically – result in some unacceptably poor behaviors when adapting to a-nominal situations is rendered impossible.

Failure to recognize how the world is changing outside the own domain, and how inevitable and serious its impact will be, is common in industrial and manufacturing automation. For instance, the Manufacturing Automation Protocol[3] failure and demise is easily explained by the insights on autocatalytic sets in Chapter 4 (i.e., caused by the absence of critical user mass).

Today, the industrial automation community needs to recognize that the sophistication of their envisaged automation projects requires the skill level of a master in information and communications technology (ICT). When considering the implications for the own career, the talented holders of such an ICT degree will be reluctant to invest their own time in industrial-automation-only technologies (as they have plenty of other opportunities to dedicate their career to).

In other words, user requirements are valid information but need further "processing" if they are to drive developments. In this respect, XPRESS has a lot of respect for the comfort zone of the industries (manufacturing and industrial automation) but has little consideration for the manner in which the world is changing, and how mainstream ICT may impact industry.

Task-Description-Driven Manufacturing (Peschl, 2014)

XPRESS distinguishes:

- *Task-driven intelligent production equipment.* Input is a task description, goal, and boundary conditions. Output is the result of the task execution. The experton/manufactron interprets the input, executes the task, and assesses the result.
- *Workflow execution at the manufacturing execution system (MES) level.* Input is the eBOP (electronic bill of processes) and feedback from previous processing steps. Output is task descriptions for the next processing steps, possibly modifying the sequence (e.g., to perform rework). Experton/manufactron orchestration by workflow and quality managers handles workflow routing as demanded by the situation and optimizing as indicated.

[3]http://en.wikipedia.org/wiki/Manufacturing_Automation_Protocol.

- *Workflow generation at the ERP level.* Input is product description and optimization target. Output is the eBOP for the MES level and task descriptions for all processing steps.

The overall idea is a task-driven manufacturing system in which expertons/manufactrons are self-contained entities, encapsulating expertise and functionality, which interact with their environment by the exchange of standardized synchronous messages. These self-contained entities aim for plug & produce.

Note that from a PROSA and ARTI perspective, the overall idea is more accurately described by "task description driven." Indeed, a key distinction from PROSA and ARTI (or Production 2000+) is the passive role of the activity instances (or work pieces). XPRESS automates the common organization in manufacturing: production lines comprising processing units/stations, which are intelligent (experts), executing tasks on dumb products/parts.

XPRESS uses TDDs (task description documents) and QRDs (quality result documents), which corresponds to holons (agents) that are so-called first-class citizens in PROSA and ARTI (Production 2000+). Resource instances have self-description documents whereas ARTI has holons.

Discussion

A major merit of XPRESS is its respect for the manufacturing community and its comfort zone. Its main weakness is too much respect for the manufacturing automation community and its comfort zone; there is a lack of attention to the influence and impact from changes beyond this community. As the wording "experton" betrays, the project sees the world through the eyes of manufacturing process experts while the remainder aims at fitting as good as possible with current practices in existing manufacturing organizations.

The innovations that XPRESS attempts to introduce (will) reside, most likely, inside the valley of death discussed in Section 7.3. The *description documents* in XPRESS may either need a level of expressiveness that requires an unachievable critical user mass or may be overly simplistic for the project's ambitions. Here, the original terminology (experton) reveals the bias of this research effort: manufacturing process experts running the factory while management remains an afterthought. XPRESS is focusing on the considerable challenges of getting the manufacturing processes running correctly and efficiently while the logistic concerns have to follow.

ARTI aims for a separation of concerns, which enables these process experts to focus on their manufacturing process without bothering with the routing, logistics, etc. But D4U also induces these experts to minimize their interfering and meddling with these workflow management concerns.

However, this does not disqualify the XPRESS results. It only positions them where most of the industrial automation is likely to end: they will implement the elementary processing steps in a manufacturing environment. However, the orchestration in a demanding and dynamic environment and integration on a larger scale will escape them. Indeed, as soon as the application's complexity escapes the abilities of a bachelor in industrial automation, a suitably skilled master in IT (or equivalent) is required. But recruiting of such human talent will require and results in a much larger role of mainstream ICT, hardened for industrial application but not invented by industrial automation (cf. industrial implementations of Ethernet).

The days in which industrial and manufacturing automation is allowed to create its own ICT solutions are gone. In communication technology, Ethernet has pushed MAP (manufacturing automation protocol) into oblivion already some decades ago. Other ICT technologies will follow. Autocatalytic set dominance (critical user mass) is a law of the artificial that cannot be ignored as soon as the conditions for its applicability become a reality. This is happening now.

Relative to D4U, XPRESS neglected to exploit all major sources for components and systems that comply with the first design principle. It has no first-class citizenship for activity instances and types; they are reduced to documents. This has repercussions for the (non-)adaptation of the technology (cf. valley of death). This impairs its range of scalability and applicability. When facing a network of production departments within a factory, organizations without production lines, and products with challenging processing requirements, obstacles will be encountered.

PROSA SIBLINGS – ADACOR (ADACOR, home page; Leitão, 2004; Leitão and Restivo, 2006)

This section discusses ADACOR – an agile and adaptive holonic architecture for manufacturing control – which is PROSA compliant but not D4U compliant.

ADACOR distinguishes four basic holons: supervisor, product, task, and operational. Product, task, and operational holons readily map onto the base holons in PROSA. However, the supervisor holons are no staff holons: they are mandatory. In PROSA terminology, this holon is an aggregated resource holon providing supervisor functionality and comprising operational holons and even task holons as sub-holons.

ADACOR recognizes two alternative states: a stationary state, where the supervisor holon provides global optimization of the production process, and a transient state, triggered by the occurrence of disturbances, where the operational holons operate in a heterarchical mode. In stationary state, the holons are organized in a federated architecture, with the supervisor holon elaborating optimized schedules as a coordinator. If the system deviates from the planned behavior, for example, due to a machine failure, the control system enters in the transient state. ADACOR provides mechanisms to return from a transient state to a stationary state. To this end, these mechanisms spread a digital pheromone establishing an "autonomy" level, which determines whether and where the control operates in a "stationary" or "transient" state.

The focus of ADACOR architecture is the shop floor level, and especially flexible manufacturing systems organized in job shop production type, characterized by concurrent and asynchronous processes with non-preemptive operations and offering alternative product routings. In this respect, ADACOR is a partial instantiation of PROSA. Conceptually, it starts from the reference architecture and introduces additional properties, reducing the scope but increasing the functionality. For its focus, ADACOR is closer to a fully operational production control system (less implementation work remains to be done). But it pays a price: outside its focus, it needs to be adapted (design choices need to be undone) whereas PROSA only needs to be further elaborated. This increased focus is not D4U compliant.

The exploration and intention propagation DMASs in PROSA and ARTI employ a different mechanism to cooperate with optimizing/scheduling holons. Each activity instance holon re-evaluates individually – when creating an intention ant – whether to follow the scheduling advice. Each resource instance holon gives priority to reservation requests in accordance with the scheduling advice. The scheduling remains optional, and switching between "stationary" and "transient" is fine-grained. Note that implementations of these individual decisions by activity and/or resource instances are able to coordinate, for example, by using a DMAS (i.e., D4U keeps options open). It is possible to have an ADACOR-like design to manage schedule adherence.

Importantly, ADACOR does not separate the generation of predictions from the scheduling for optimization purposes. PROSA and ARTI acknowledge that predictions are useful on their own (e.g., to organize auxiliary operations). Moreover, this D4U prediction generation – as it is unconcerned with and therefore unconstrained by the needs of optimization mechanisms – is able to give priority to an accurate reality-reflection covering all the might-be-relevant concerns and characteristics. This distinctive quality of D4U designs (PROSA, ARTI, DMAS) is often overlooked in the discussion of work-by-others by researchers aiming for optimized production. In addition, the manner in which activity instances reconnect with scheduling advice in a fine-grained manner equally fails to be registered in these discussions. The ability to adopt schemes for reconnection with advice, used by others, remains completely out-of-sight.

ADACOR is a prominent member of this collection of "PROSA siblings." ORCA-FMS is another example (Pach et al., 2014). Here, activity instance holons never return to a stationary mode. However, they switch individually. The idea is that the ORCA-FMS applicability range is characterized by relatively short-lasting production times (from product launch till shipment). Therefore, switching back is not important when more recently launched (or to-be-launched) products benefit from a schedule based on up-to-date information.

Relative to the work discussed in this book, this work-by-others spends considerably more effort on performance optimization and evaluation. This includes research into novel optimization mechanisms that adapt, self-organize, etc. Until now, such performance optimization remains out of reach for D4U, as it is inherently choice-rich and noncompliant with the first design principle. How much is inherently and unavoidably out of reach for D4U remains an open question. The prediction capabilities, generated by intention ants, also seemed out of reach in the past. It stays undecided whether there are more discoveries to be made reconciling D4U with performance. However, these are likely to bring a body of knowledge or a collection of mechanisms/templates, not an enhanced architectural feature or pattern.

ANT COLONIES AND STIGMERGY

This section presents work by others that is mistakenly considered to be closely related to D4U research. Two examples are discussed. There exists undoubtedly other work that is erroneously assumed to be relevant in a D4U context.

ACO – Ant Colony Optimization

ACO is a single-shot problem-solving mechanism. This category of research is completely out of scope for the research in this book. Holonic execution systems address going concerns.

This distinction is fundamental (Wegner, 1997). Taken to its extreme, a technology like IBM's Watson is outclassed by a basic thermostat in view of Wegner's insight that interaction is more powerful/expressive than computation. Watson receives a query, goes to work, and delivers a result. Nothing happens until Watson receives another query. Each query initiates a single-shot problem-solving activity. The thermostat takes care of a going concern: keeping the temperature at its sensor close to the setting. It interacts with its environment and it takes care.

It makes little sense to discuss or address single-shot problem solving and handling going concerns in a single research context. Single-shot problem solving allows to measure, qualify, and compare performance to an extent that is unachievable for research addressing going concerns. Going concerns research may utilize single-shot problem solving as a step in an ongoing problem handling process. But cooperation across the border between the two classes is pointless, even counterproductive, as soon as one party fails (refuses) to appreciate how different these classes are (analogy: "still photography specialists" looking at real-time TV images and judging based on the quality of single images out of a TV broadcast fragment).

Stigmergy in Agent-Based Simulation and Decision Support Systems

Another class of applications, which was inspired by stigmergy in ant colony food foraging, develops considerably simpler systems than the holonic execution systems discussed above. The world of interest is typically modelled as a 2D grid, possibly with some levels (reflecting multistory buildings) or a shaped surface (reflecting a mountainous area).

In this grid model, simple agents (e.g., programmed as small state machines) evolve, deposit simple digital pheromones (i.e., scalar values and the pheromone type identifier), and sense digital pheromones. Importantly, these applications are sufficiently simple for one person to be able to program them in a matter of days, maybe a few weeks. The challenge is in tuning these models and designing the agents (state models). Parunak and Brueckner have developed numerous applications based on this approach.

An interesting example utilizes automated programming to generate the agent programs (Parunak et al., 2007). The world model is extended with a discrete third dimension: time. It is like a book with the middle page corresponding to the present, the previous pages to the past, and the following pages to a – predicted – future. The past pages contain the observed states (trace). The automatically generated agents are injected in the past and execute their – programmed – behavior until the present. A criterion is evaluated to estimate how well they fit the past. The best-performing agents are allowed to progress into the future. The result is used for decision support.

Note how this solution can operate online, processing new information as it becomes available. It addresses the going concern of providing a predictive situational awareness in situations where knowledge about the world of interest is lacking and/or less than perfect. It not only predicts but learns about the characteristics of the underlying system in a manner that can be used for the predictions.

This is an approach that can be integrated in an ARTI system. Machine learning can be used to estimate relevant properties, preferably with lots of training data and slow changing properties. The learned properties can then be used for virtual execution, or DMASs that deposit these scalar digital pheromones can be used. In the latter case, the simplicity – which is very desirable – normally introduces a tuning task: how to interpret these scalar values that have no direct relationship to the corresponding reality. Overall, this is complementary and captivating research. It shares interests with D4U, unlike single-shot problem solving.

ABBREVIATION

ACO Ant colony optimization

REFERENCES

ADACOR, Web page: www.ipb.pt/~pleitao/index.php/adacor.

Bussmann, S., Jennings, N.R., Wooldridge, M., 2004. Multi-agent systems for manufacturing control: a design methodology. In: Series on Agent Technology. Springer-Verlag, Berlin, Germany.

Pach, C., Berger, T., Bonte, T., Trentesaux, D., 2014. ORCA-FMS: a dynamic architecture for the optimized and reactive control of flexible manufacturing scheduling. Comput. Ind. 65(4), 706–720.

Parunak, H.V.D., Brueckner, S., Matthews, R., Sauter, J., Brophy S., 2007. Real-time evolutionary agent characterization and prediction. In: Sixth International Joint Conference on Autonomous Agents and Multi-Agent Systems, Industrial Track.

Leitão, P., 2004. An agile and adaptive holonic architecture for manufacturing control, PhD thesis, University of Porto, www.ipb.pt/~pleitao/pjl-tese.pdf.

Leitão, P., Restivo, F., 2006. ADACOR: a holonic architecture for agile and adaptive manufacturing control. Comput. Ind. 57(2), 121–130.

Peschl, M., 2014. An architecture for flexible manufacturing systems based on task-driven agents. PhD thesis, University of Oulu, jultika.oulu.fi/Record/isbn978-952-62-0366-9.

Schild, K., Bussmann, S., 2007. Self-organization in manufacturing operations. In: Communications of the ACM, Vol. 50, No. 12, pp. 74–79, Berlin, Germany.

Wegner, P., 1997. Why interaction is more powerful than algorithms. In: Communications of the ACM, Vol. 40, No. 5, pp. 80–91, New York, USA.

XPRESS, Flexible production experts for reconfigurable assembly technology. FP6-NMP Project no. 26674, 2007-2011.

CHAPTER NINE

Summary and Outlook

SUMMARY

The story behind this book started by a fairly minimalistic modeling of the mechanisms causing the failure of system integration efforts. From the insights provided by this theoretical model, the research looked into the design of systems and components avoiding the failure-causing mechanisms. This resulted in *design for the unexpected*. Indeed, when developers want to ensure that their design is suited for integration without control over what needs to be integrated with, everything – even the unexpected – needs to be accounted for (Taleb, 2010).

Design for the unexpected is not a brute-force approach, vainly attempting to account for all eventualities. Instead and first, it is about developing as much as possible elements of a solution without the need to rely on expectations (Chapter 3: Design Principles). Subsequent developments that cannot avoid relying on expectations are introduced in manners that still minimize this relying on expectations and minimize the effort required to revise or undo. In other words, *this book is about a scientific approach to low-and-late commitment* (while still allowing for early preparation).

Inherently, *design for the unexpected only provides an intermediate solution* – a platform or basis – on which final solutions still have to be elaborated (see Section "The Watchmakers' Parable" of Chapter 4). The (stable) intermediate solution reduces the time and effort needed for such finalization while improving service levels considerably. The surprising aspect of this research is how much can be achieved by designing for the unexpected first and how little remains to be done by finalization efforts afterward.

The most advanced research result is the development, based on the Design Principles (Chapter 3) and the Laws of the Artificial (Chapter 4), of *a reference architecture* (PROSA and its generalization ARTI) enabling the description of the structure of the complex adaptive systems occurring and emerging in mechatronic societies, and their interactions. The formalism is scalable by its *rigorous "separation of concerns"* (Section "The PROSA Reference Architecture" of Chapter 5).

Design for the Unexpected. http://dx.doi.org/10.1016/B978-0-12-803662-4.00009-6

The control dynamics (task execution) are ruled by the DMAS (Section "Bio-inspired Coordination and Control in Holonic Execution Systems" of Chapter 5), with the *ability of short-term forecasting* as the most important ensuing feature, obtained by a *decentralized virtual execution* based on a *digital mirror image of the world of interest* that reflects reality at all times, as a *single source of truth*. DMAS is scalable and provides robustness to the controlled system.

The universality and general applicability of the PROSA/ARTI/DMAS framework are convincingly demonstrated in Chapter 7, with case studies from a wide variety of mechatronic societies.

OUTLOOK – TOWARD A HUMANE AND RESPECTFUL MECHATRONIC SOCIETY

Today's society is transforming into a mechatronic society at a breath-taking pace: ubiquitous computing, ambient intelligence, web 2.0, Internet of Things, big data analytics, smart power grids, smart homes, smart cities, smart harbors, intelligent traffic, intelligent multimodal transport and logistics, intelligent networked manufacturing, intelligent products, eHealth, domestic robots, smart watches, self-driving cars, industry 4.0, etc. There is no shortage of such buzzwords and slogans corresponding to the creation of a world in which information processing and communication increasingly play a larger role.

Every major change in society, although initiated to reap benefits, creates issues. Privacy, responsibility, accountability, empowerment, social control, etc. all need to be reinvented (Floridi, 2015). In today's initial stages of this transformation of our society, much remains to be learned. Among others, the increasing returns[1] characterizing ICT naturally result in monopolistic and oligopolistic situations (e.g., Microsoft, Google, and Apple). Large organizations, including governments and other authorities, obtain the leverage and power to unhealthy, risky[2] extents. Whenever these large organizations experience the wrong kind of pressure, which only is a matter of time, abuse and damage to society will become reality.

[1]https://hbr.org/1996/07/increasing-returns-and-the-new-world-of-business

[2]The self-organizing mechanisms in our society imply that overly powerful authorities, where this amount of power often is justified by crime fighting requirements, result in organized crime becoming the authorities. Indeed, there is no balance/divide of power to control the authorities and there is no way to hide, for organised crime, except by becoming these authorities. It is only a matter of time and not a question whether this will happen.

The contribution of design for the unexpected in this respect comprises an architecture that stays close to its world of interest. D4U solutions mirror the corresponding reality precisely to be protected against unexpected future demands (Chapter 6). This has a beneficial side effect: humans are mirrored in a single-source-of-truth design. For every human, there is a single "e-person" available 24/7. In a D4U design, this e-person provides the maximum amount of services and functionality related to the corresponding real-world person. Indeed, this is exactly what makes the D4U solution scale-able, integrate-able, and able to survive unexpected demands.

This opens perspectives to empower the individual well beyond the current state of affairs. Information requests related to a person can be handled by this e-person, whereby the human stays in control. For instance, requests for medical information can be honored by not only providing the requested information but also volunteering relevant additional information (e.g., flagging a relevant risk factor related to the requested information). When all such requests are directed at this e-person, the corresponding human has a handle to self-manage information processing that concerns his or her person in cyberspace. Among others, the e-person will have an overview of the parties asking and providing information.

This empowerment through an e-person provides, at least conceptually, solutions to tricky issues. For instance, when a powerful organization makes an unauthorized request (e.g., for sensitive medical information), this e-person may provide sanitized information. When another person (e.g., the medical doctor of this person) receives a similar request for the same information from this unauthorized but powerful organization, consistently sanitized information may be provided as this other person (medical doctor) consults the e-person (of the patient) concerned. This effectively neutralizes the dominant position that a powerful organization may abuse (e.g., under competitive pressure).

Another example is an e-person receiving a request to confirm whether some information is true and this is about an embarrassing situation some 20 years ago during this person's youth. The e-person may label it as "please forget, I sincerely regret..." and receivers of information about other persons may filter such information out.

Such innovative solutions transcend the technological aspects on which this book focuses. They call for inter-, trans-, and multidisciplinary approaches in which, for instance, an effective strategy is elaborated to establish an appropriate kind of "political correctness." This correctness involves always consulting the e-person and respecting its wishes in the absence of ample, relevant justification to do otherwise. Likewise, legal measures need elaborating to achieve desired behavior under heavy pressures (e.g., in competitive or political domains). Here, D4U needs joint research and development with the social sciences and humanities.

Moreover, D4U offers (the potential to establish) superior services, in comparison to current practice, precisely because it empowers its users (both individual persons and groups). In particular, predicting the unexpected (by exploring and intention propagation DMAS designs) delivers a service that is out of reach of, for example, big data analytics. Indeed, the best way to predict the future is to create the future. The intention propagation in combination with commitment mechanisms is exactly implementing this saying (in a distributed and collective manner). Here, persons are in charge of their future (subject to respecting others) when their e-person searches for and commits to a future by using DMASs on an ARTI infrastructure.

Again, this transcends technology; social sciences and humanities become relevant, essential factors. Legal support may be needed to enforce (self-decided, self-imposed) commitments. For example, a person leaving for work late, allowing others to drive congestion-free to work during rush hours, may have a commitment from the other drivers to leave a given parking space free. This must be enforceable (such good deeds must remain unpunished if a superior service from D4U traffic management systems is to become a reality). Likewise, policy design needs to address situations in which the individuals cannot agree within the DMAS-generated predictions (e.g., what happens when the prediction indicates an undesirable congested state and no improvement can be observed).

Moreover, completely new models for a world of interest may need investigating. For instance, how can we model a world of interest for an innovative change management in organizations that empowers the individuals concerned, both for radical changes and continuous improvement processes? Current practice reveals huge margins for improvement (cf. statistical data on burnouts and depressions). How can we achieve the value-creation, characteristic for market based organizations, in public services? How can we counter unhealthy dominance in supply chains, resulting in selective and poor information flows in markets (e.g., failure to have increasingly better food at constant prices instead of apparently constant, but de facto lower, quality food at lower prices)?

Today, the main topic on our research agenda is to create a mechatronic society that is empowering, considerate, and hospitable toward humans. This book reveals the tremendous opportunities that are still available and partially uncharted to make better use of the world's scarce resources, to a better functioning of the world's infrastructure at all levels of size and complexity, and to a more harmonic (mechatronic) society. Here, the answers and solutions are available to researchers that are willing, able, and allowed (by society) to transcend the borders of their own discipline and comfort zone.

REFERENCES

Taleb, N., 2010. The Black Swan, second ed. Random House, New York.
Floridi, L. (Ed.), 2015. The Onlife Manifesto, Being Human in a Hyperconnected Era. Springer International Publishing, ISBN 978-3-319-04092-9.

What are (Software) Agents?

There is no widely accepted definition of the term agent that the authors are aware of, which – by definition[1] – implies that there exists no such definition. A weak definition of a computing agent given by Wooldridge & Jennings (Wooldridge, 2002).

Definition: An agent is *a software program* with the following properties:

- *Autonomy*: agents operate without the direct intervention of humans or others, and have some kind of control over their actions and internal state ...;
- *Social ability*: agents interact with other agents (and possibly humans) via some kind of agent-communication language ...;
- *Reactivity*: agents perceive their environment (...), and respond in a timely fashion to changes that occur in it;
- *Proactiveness*: agents do not simply act in response to their environment; they are able to exhibit goal-directed behavior by taking the initiative.

Note that strong(er) definitions of an agent mainly reflect research communities attempting to "hijack" something that may strengthen their "brand name" in order to get access to funding, prestige, etc. and, therefore, rarely provide any significant added value.

D4U, PROSA, ARTI, and DMAS mostly correspond to situations in which several agents are interacting with each other, which leads to the following definition:

Definition: A multiagent system (MAS) is a set of interacting agents.

DESIGN FOCUS

When developing such a multiagent system, two aspects have to be considered:

- *Agent design*, focusing on the design of (the internal model of) the individual agent.
- *Society design*, focusing on the responsibilities and corresponding interaction between the different agents without specifying the internals of an agent.

[1]To be widely accepted, a term needs to be widely known.

Design for the Unexpected. http://dx.doi.org/10.1016/B978-0-12-803662-4.00010-2

The BDI (Beliefs-desires-Intentions) model (Rao and Georgeff, 1995) and the subsumption architecture (Brooks, 1986) are both concerned with agent design, as they offer guidelines for the design of an individual agent. The PROSA and ARTI reference architectures on the other hand can be considered as involved with society design.

SITUATED IN AN ENVIRONMENT

When using agent technology in a holonic execution system, D4U, PROSA, ARTI, and DMAS have their agents evolve in a digital image of the relevant reality: *the agents are situated in a software environment.* In contrast to high-profile research on situated agents, D4U agents may reason by performing sophisticated computations and communications. What is important is that these situated agents do not need their own representation of the world to perform this reasoning. For example, each relevant physical entity has a software counterpart in this software environment (the digital image). There will be a single-source-of-truth design, which makes it easier to maintain a correct and up-to-date representation of the world of interest and it avoids inconsistencies and incompatibilities between internal representations of different agents.

REFERENCES

Wooldridge, M., 2002. An Introduction to MultiAgent Systems. John Wiley & Sons, Chicester, UK.

Rao, A.S., Georgeff, M. P., 1995. BDI-agents: from theory to practice (PDF). Proceedings of the First International Conference on Multiagent Systems (ICMAS'95).

Brooks, R.A., 1986. A robust layered control system for a mobile robot. IEEE J. Rob. Autom. RA-2, 14–23.

APPENDIX II

Simulation, Emulation, and Modeling

The innovative software platform designs discussed in this book utilize simulation in a multitude of places for different purposes. However, the words *simulation* and *emulation* have widely varying meanings depending on the context in which they are used. This appendix intends to clarify this issue for the purposes of the discussion in this book.

SIMULATION – WHAT IT IS NOT (FOR US)

This section concisely discusses simulation as it is used/defined by others but not as the present discussion considers relevant. It is a nonexhaustive list:

- Animation. In many domains, for example, entertainment, robotics, or machining, a real-world activity is simulated in 3D. The software generates a movie that visually depicts/simulates the object of the simulation. For instance, a pick-and-place operation by a robot.
- Physics. A whole subdomain of simulation targets physical systems such as computational fluid dynamics, targeting the flow of air over a wing of a next-generation aircraft.
- DEDS. In discrete event dynamic systems, mathematical models are used to analyze and simulate. The key requirement for the models is to allow and support analysis techniques; to this end significant simplification of reality commonly needs introducing. The theoretical nature of these models makes it nontrivial to ensure a correct correspondence between reality and the model, failing to guarantee the ability to mirror the structure of reality in the model (i.e., composition-ability).
- KISS. *Keep it simple s…* simulation is the prevailing textbook approach to simulation. Here, a simulation model of the complete system – that is to be simulated – is modeled keeping the simulation as simple as possible. In others words, it will be a simulation targeted and limited to answering specific questions (e.g., how long it will take to empty a car park near a football stadium). Again, such simulations are unlikely to mirror the structure of what is simulated from reality.

Design for the Unexpected. http://dx.doi.org/10.1016/B978-0-12-803662-4.00011-4

There are two motivations for excluding the above. The first reason is overkill (the simulation goes into too much detail). The required extra effort will cause the simulation to be time-consuming while the extra information has little or no importance. Such simulation still may contribute indirectly. For instance, it can be used to compute (initial values for) parameters of the models below (e.g., to compute the duration of a pick-and-place operation by a robot) and/or initial parameter values for components that will use on-line estimation methods (e.g., machine learning to track the duration of this pick-and-place operation). In addition, these more detailed simulations might be used during ramp-up and deployment. They can be used to confront the coordination and control system with an almost real-world challenge and allow to remedy issues before establishing the connection to the real world (and exposing real-world assets to teething problems).

The second reason is too large a distance between the world-of-interest that is simulated and the model (elements). This prevents composition of the models when good-enough specific-purpose models are used. More importantly, this causes validation issues. When a model of a conveyor models directly how it behaves, measurements on such a conveyor provide the necessary data. When an aggregated model simulates a transport system, statistical validation is required where the range of conditions, under which the validation holds, remains difficult to assess. This kind of simulation lacks reusability and compose-ability. This type of simulation has little relevance for our purposes but shares simulation techniques (i.e., they use more or less the same tools and programming but apply a different methodology: fast results but not reusable).

SIMULATION – WHAT IT IS (FOR US)

Simulation and Emulation

The present discussion is concerned with simulation models that render interacting with the real world indistinguishable from interacting with a (model-using) emulation for the coordination and control. The coordination and control system may dispatch an operation and receive feedback on its progress and outcome from a simulation model or from an interface to a corresponding real-world operation. It will not perceive a difference. For instance, when a robotic pick-and-place is initiated, the simulation will send information as if the real-world operation took place.

In this case, it is possible to call this emulation: the simulation mimics/replaces a real-world component or subsystem. The rest of the world is unaware and unexposed to the fact that it is interacting with a computer program instead of a real-world process.

A key advantage/requirement of our approach is to mirror the structure/composition of the world that is simulated. This enables an initial validation on a component per component basis. For example, the emulation of the robot operations need to generate the right delays (time to perform a given operation) and outcome distributions (e.g., a small percentage of failures).

Software in the Loop

In order to validate the simulation of the coordination and control software automatically, this software will be included in the simulation without modification except for an additional capability that can be disabled in deployed systems. This software in the loop simulation ensures that validation of the simulation of the coordination and control is automatic and without any effort.

The additional capability consists of the coordination and control system being able to signal when it is idling. This will allow a significant speed-up of the simulation: the simulation may jump on its time axis to the next event. When the coordination and control system is busy, the simulation has to remain in a real-time mode such that events keep happening at the correct time (while the coordination and control is taking perhaps too much time to decide what to do).

Next to software, the hardware in the loop may be part of a deployment and ramp-up campaign. This entails that the coordination and control system is not only using the actual software within the simulation but also a computer network, as it will be deployed. Computing processes now experience more realistic delay when communicating over, for example, the Internet. However, hardware in the loop is optional and not an essential part of the approach in this book. Software in the loop is required.

Simulating Decision Making

To simulate as a standalone implementation, all decision-making processes need a model. Such standalone simulation comprises emulation models for the resources and processes in the world of interest but also executable models for the decision-making processes. Here, there are three categories:

- Simple decision-making software. This software will be its own model.
- Complex decision-making software. This may be its own model in the later phases that are part of a deployment activity (cf. hardware

in the loop and use of detailed/physical simulation models). However, there is a need for approximating simulation models, which are used for much-faster-than-real-time simulation (see further). These computationally simple models may use data generated by the complex decision-making software, which is then executed at a lower frequency.
- Human decision-making. There is a need for computationally simple models that generate – estimated – decisions made by the humans that they model. In deployed versions, these models may learn from feedback about the real decisions made by real humans.

Below, the uses of simulation are discussed in more detail. This will reveal why different decision-making models are needed, and how models are reused for different purposes.

Finally, decision-making software may need to support two or more modes:
- Low effort and good-enough estimation of the real decision-making outcome (e.g., only used to plan activities where this planning can/will be revised regularly).
- High effort producing a decision that can be executed in reality (e.g., needs to be signed off by an authorized person).

Currently, two modes are sufficient, but an intermediate version may emerge in future designs.

Hybrid Real-Time and Discrete-Event Simulation

Our simulation needs to address two conflicting requirements:
- Speed. As much simulation time needs to be covered in as little real-world time as possible. This calls for discrete event simulation in which simulation time jumps to the next event as soon as the current event has been processed. Many simulation tools use this approach assuming that decision-making takes negligible time. Unfortunately, this assumption does not hold in our cases.
- Account for the time needed for decision-making. This calls for real-time simulation in which event are generated when they occur on the real-time axis. When a decision-making process takes (too) much time, events will occur and the simulation will make the effect visible (e.g., poor performance caused by decisions after the facts).

As stated, the coordination and control software, executing in the loop, signals whether it is idle or not. When idle, the fast discrete event mode is used. Otherwise, the real-time simulation mode is used.

USES OF SIMULATION

The simulation/emulation models are used multiple times.

Embedded Simulation at Run-Time

The proactive coordination and control systems use the executable models for – at least – two purposes. First, the exploration DMAS uses simulated/virtual execution to discover and assess/evaluate candidate courses of action. Second, the intention propagation delegate MAS uses simulated/virtual execution to inform all resources of their expected loadings/usage while providing each activity with its expected routing, including timing.

Here, much-faster-than-real-time virtual execution is essential. The models generate expected behavior. Alternatively, some safety margins might be included. Advanced models may generate probability distributions instead of scalar values (= future research).

Standalone Simulation

Whereas the above use occurs during actual deployed operations, the standalone simulation connects the above to an emulated version of reality. The executable models used for virtual execution by the DMAS, rendering the system proactive, are used twice. First, they are used as above (= software in the loop). Second, the same models emulate the world of interest; the coordination and control system can/will not see any difference with the embedded mode, except that its idling indicator will be used to speed up the simulation. The embedded simulation operates in real time for obvious reasons (i.e., the events originate from the world of interest, which happens to be real).

The models emulating reality are generating stochastic data, sampled from the probability distributions in the models. The embedded models generate nonsampled distribution parameters (e.g., median values).

Deployment Supporting Simulation

Transitions from the standalone simulation toward the deployed system with embedded simulation may be performed directly. However, simulation is able to provide intermediate steps in which more control and observability can be provided while real-world assets stay out of risk.

The emulation of the real-world system can be done in more accurate – usually much slower – simulation models (often providing detailed animation). Hardware in the loop can be added, and human decision-making and complex decision mechanisms can be introduced. All this is optional.

TECHNOLOGIES USED FOR SIMULATION

Chronologically, software technologies have been used for research purposes. There has been an evolution from solutions suited for small-scale academic prototypes toward a highly scalable solution supporting emulation model development at commercially viable speed and cost. The main steps are discussed concisely:

- The initial solution was a combination of an established industrial simulation tool in its most advanced version (i.e., Arena Professional edition supporting user-defined model templates) with C++. This solution only supported real-time mode. The hybrid mode is impossible because industrial simulation software does not support switching between real-time mode and discrete-event mode within a single simulation run. In fact, most industrial tools only support discrete-event mode.
- This initial solution is complicated (e.g., needs a message distribution system in Arena cooperating with a C++ counterpart). It also requires proficient developers knowledgeable in both C++ and the Arena modeling/programming facilities. In summary, it was a barely workable solution allowing to develop academic prototypes that were impractical to maintain.
- The second solution adopted Java in combination with a simulation software library. Difficulties to debug software – where it remained unclear whether the problem resided in the library, the Java control system or the cooperation among those two – gradually reduced our reliance on the library. After a relatively short transition period, the entire development occurred in standard Java.
- During that time, both industrial software tools (e.g., FLEXSIM, IEC64199) and multiagent tools (e.g., Jade, AnyLogic) have been assessed but offered advanced functionality where it was not needed without functionality compensating for their smaller user community (relative to Java). The native Java solution enjoys a large community of users and proficient professionals. It allowed for larger prototype implementations that could be adapted to address novel problem domains. However, scalability and maintenance remained a serious concern.
- In 2011, the team adopted Erlang/OTP (cf. www.erlang.org). This solved the scalability issue. Model development in Erlang enjoys the speed-up brought by a functional/actor language with pattern matching (one or two orders of magnitude faster development). Other advantages are high availability, hot code updating, and distributed computing. It also connects excellently to other languages and can run on embedded computers (e.g., raspberry pi). Software model maintenance and sharing

within a community is subject of ongoing research. Widespread acceptance of the technology still remains undecided (but multicore computers are pushing us in this direction).

- One Java implementation, which had to be connected to a .net system programmed in C#, was ported to the Erlang solution in a couple of weeks while the connection to C# occurred in a single day. Erlang / OTP is a product from the telecom industry.

THE HYBRID SIMULATION ENGINE

The hybrid – switching between real-time and discrete-event modes – simulation engine comprises two Erlang modules: event_table and event_mgr. The event_table contains all the events that have been created by the holons. The manager generates the next event in the table, either shortly after sufficient time has passed (in real-time mode) or when entering the discrete-event mode. The code below is not optimized for performance but rather for readability and observability. Readers familiar with Erlang may inspect it to understand how this hybrid simulation works. In general, the code provides an indication how (not so) complex this simulation approach is.

```
-module(event_table).

-export([init/0, insert/1, do_first/0, nextEventTime/0]).

init() ->
      ...
      ets:new(event_ets_table, [ordered_set, public, named_table]),
      insert({infinity, eventmgr, stop, []}),
      {ok, event_ets_table}.

insert({Time, M, F, Args}) ->
            ets:insert_new(event_ets_table, {{Time, make_ref()}, M, F, Args}).

do_first() ->
      Key = ets:first(event_ets_table),
      [{ Key , M, F, Args}] = ets:lookup(event_ets_table, Key),
      ets:delete(event_ets_table, Key),
      spawn(M, F, [ Key | Args] ).

nextEventTime() ->
      {Time, _ } = ets:first(event_ets_table),
      Time.
```

```
% Explanations:
%
% Please look at the Erlang/OTP documentation
% for the ets module used in the sample code.
%
% The {Time, M, F, Args} parameter is composed of:
% - Time of the event to be generated by the event_mgr (in milliseconds)
% - M is the module selected (and typically implemented)
%   by the poster of the event.
% - F is the function exported by module M selected by
%   the poster of the event.
% - Args is the list of arguments to be given to F when
%   do_first() creates a process that executes F.
%   Args will be extended by a time_token (i.e. {Time, make_ref()}).
-module(eventmgr).

%% user functions
-export([start/0, post_abs/1, post_incr/1, return_token/1, get_time_token/0]).
-export([stop/0, stop/1]).
%% for internal use only
-export([loop_rt/4, loop_des/2]).

start() ->
      event_table:init(), %% create the event_ets_table
      register(event_manager,
                  spawn (eventmgr, loop_rt, [0, infinity, 0, {0, now()}])),
      ok.

stop() -> event_manager ! stop.
stop( _ ) -> stop(). % when called from event_table:do_first()

% post an event with an absolute time coordinate.
post_abs({Time, M, F, Args}) ->
      event_manager ! {post_abs_event, {Time, M, F, Args}}, ok.

% post an event with a relative/incremental time coordinate.
post_incr({Delay, M, F, Args}) ->
      event_manager ! {post_incr_event, {Delay, M, F, Args}}, ok.

% return a token to indicate all event processing is done.
return_token(Token) ->
      event_manager ! {return_a_token, Token}, ok.

% get simulation time and receive a token ensuring switch to real-time mode.
get_time_token() ->
      event_manager ! {get_time_and_token, self(),R = make_ref()},
      receive
            {R, {Time,Token}} -> {ok, {Time, Token}}
      end.
```

```
%% loop_des(                  discrete event mode (des mode)
%%        NextEventTime,      Time for the earliest event in event_ets_table
%%        SimulationTime,     Current simulation time in des mode
%%       )

loop_des(infinity, SimTime) ->
      %% switching to real time mode when no events are pending
      loop_rt(0, infinity, SimTime, {SimTime, now()});

loop_des(NextEventTime, SimTime) when NextEventTime > SimTime ->
      %% In des mode, jumping to the time of the earliest event
      loop_des(NextEventTime, NextEventTime);

loop_des(NextEventTime, SimTime) when NextEventTime =< SimTime ->
      %% Simulation time has caught up with the earliest "posted event".
      event_table:do_first(), %% issues a time_token; switching to rt mode
      loop_rt(1, event_table:nextEventTime(), SimTime, {SimTime, now()}).



%% Compute simulation time while in real time mode.
adaptTime({RTentryTime, ClockAtRTentry}) ->
      %% RTentryTime is value of the simulation time at entry in RT mode.
      %% ClockAtRTentry is the return value of now() at entry in RT mode.
      RTentryTime + (timer:now_diff(now(), ClockAtRTentry) div 1000).
%% loop_rt(                   real-time mode
%%       (
%%        TokenCount >= 0,    Number of tokens to be returned for DES mode
%%        NextEventTime,      Time for the earliest event in event_ets_table
%%        SimTime,            Current simulation time when entering loop_rt
%%        Clock               Information to compute SimulationTime in rt mode
%%       )

%% event_ets_table is empty, awaiting event postings in real time mode
loop_rt(Count, infinity, _ , Clock) ->
      receive
            {post_incr_event, {Delay, M, F, Args}} ->
                  CurrentTime = adaptTime(Clock),
                  event_table:insert({CurrentTime + Delay , M, F, Args}),
                  loop_rt(Count, event_table:nextEventTime(),
                                                    CurrentTime, Clock);
            {post_abs_event, {Time, M, F, Args}} ->
                  CurrentTime = adaptTime(Clock),
                  event_table:insert({Time, M, F, Args}),
                  loop_rt(Count, event_table:nextEventTime(),
                                                    CurrentTime, Clock);
```

```
        {get_time_and_token, ReplyPid, R} ->
            CurrentTime = adaptTime(Clock),
            ReplyPid ! {R, { CurrentTime, make_ref() }},
            loop_rt((Count + 1), event_table:nextEventTime(),
                                                CurrentTime, Clock);
        stop -> void
    end;

%% Simulation time has caught up with the earliest "posted event"
loop_rt(Count, NextEventTime, SimTime, Clock) when NextEventTime =< SimTime ->
    event_table:do_first(), %% issues a time_token; remaining in rt mode
    loop_rt((Count + 1), event_table:nextEventTime(),
                                            adaptTime(Clock), Clock);
%% all tokens are back, switch to DES mode
loop_rt(0, NextEventTime, _ , _ ) ->
    loop_des(NextEventTime, NextEventTime);

%% waiting in real-time mode for event posting or until “next event time”
loop_rt(Count, NextEventTime, SimTime, Clock) ->
    WaitingTime = min(NextEventTime - SimTime, 1000000000),
    receive
        {post_incr_event, {Delay, M, F, Args}} ->
            CurrentTime = adaptTime(Clock),
            event_table:insert({CurrentTime + Delay , M, F, Args}),
            loop_rt(Count, event_table:nextEventTime(),
                                                CurrentTime, Clock);
        {post_abs_event, {Time, M, F, Args}} ->
            CurrentTime = adaptTime(Clock),
            event_table:insert({Time, M, F, Args}),
            loop_rt(Count, event_table:nextEventTime(),
                                                CurrentTime, Clock);
        {return_a_token, _ } ->
            loop_rt((Count - 1), event_table:nextEventTime(),
                                            adaptTime(Clock), Clock);
        {get_time_and_token, ReplyPid, R} ->
            CurrentTime = adaptTime(Clock),
            ReplyPid ! {R, { CurrentTime, make_ref() }},
            loop_rt((Count + 1), event_table:nextEventTime(),
                                                CurrentTime, Clock);
        stop -> void
        after (WaitingTime) ->
            loop_rt(Count, event_table:nextEventTime(),
                                            adaptTime(Clock), Clock)
    end.
```

APPENDIX III

Design by Abduction No Longer Suffices

When a community designs a complex system or infrastructure, abductive reasoning is commonly used to discover and identify candidate solutions. Unfortunately, it also is used to justify the solution that will be adopted as the only option (as if they had used deduction to arrive at the selected design).

Abduction is a kind of logical inference that could be described as educated guessing. When C is assumed to be true and C follows from A, abduction puts forward the hypothesis that A is true. This clearly is a search heuristic rather than a logical deduction. Indeed, logical inference only yields that $\neg$C implies $\neg$A. In fact, there is no evidence regarding the nonexistence of alternative hypotheses $A_1, A_2, \ldots A_n$ that equally are sufficient but not necessary conditions for C.

When C is the service or functionality required from a system or infrastructure, the community involved will use abduction to generate valid candidate designs $A_1, A_2, \ldots A_n$. In other words, implementing any design $A_x \in \{A_1, A_2, \ldots, A_n\}$ will result in a world in which the services required by C will be available. Next, standard practice in abductive reasoning searches for and selects the most economical design A_x among the candidates. This last procedure is used to justify calling A_x *the* solution whereas, logically, A_x is only *a* solution (among many). Such a claim that the choice of solution A_x is "beyond discussion" is unjustified in a fast-changing and unpredictable world.

Indeed, this most economical criterion favors the quick and dirty solution that is good enough. Moreover, there often exists another criterion that filters the set of candidate solutions before the most economical criterion is applied: the comfort zone. Candidate solutions that are outside the comfort zone of the decision makers in the community that controls the development and deployment of the solution will not be generated nor will they be retained in the set of candidate solutions. Especially when going outside this comfort zone may reduce the dominance of these decision makers over their domain, filtering will be stringently applied. This regularly results in a very poor service at high costs in comparison to what is

Design for the Unexpected. http://dx.doi.org/10.1016/B978-0-12-803662-4.00012-6

inherently achievable. Moreover, it reinforces the existing lock-in by legacy designs, imposing severe limitations.

Design for the unexpected prefers candidate solutions that are the least likely to cause future conflicts, both with future requirements and other solutions, especially solutions that were designed for the unexpected. Formally, it prefers partial solutions (insufficient to guarantee C) that have no conflict with anything that has no conflict with C or reality. In other words, when an unexpected demand emerges that respects C and reality, these partial solutions remain intact and useful.

Background information can be found at butte.edu/departments/cas/tipsheets/thinking/reasoning.html, which classifies and typifies reasoning as follows:

- Deductive reasoning: conclusion guaranteed
- Inductive reasoning: conclusion merely likely
- Abductive reasoning: taking your best shot

SUBJECT INDEX

H

I

T

U

V

W

X

FSC
www.fsc.org
MIX
Papier | Fördert gute Waldnutzung
FSC® C083411

Kolibri 360 GmbH
Ferdinand-Jühlke-Straße 7
99095 Erfurt, Deutschland
produktsicherheit@kolibri360.de